Guide to the FIDIC Conditions of Contract for Construction

The Red Book 2017

Michael D. Robinson

WILEY Blackwell

This edition first published 2023
© 2023 John Wiley & Sons Ltd

The right of Michael D. Robinson to be identified as the author of this work has been asserted in accordance with law.

Registered Office
John Wiley & Sons Ltd, The Atrium, Southern Gate, Chichester, West Sussex, PO19 8SQ, UK

Editorial Office
9600 Garsington Road, Oxford, OX4 2DQ, UK

For details of our global editorial offices, customer services, and more information about Wiley products visit us at www.wiley.com.

Wiley also publishes its books in a variety of electronic formats and by print-on-demand. Some content that appears in standard print versions of this book may not be available in other formats.

Library of Congress Cataloging-in-Publication data is applied for

Hardback ISBN: 9781119856627

Cover Design by Wiley
Cover Image: © Samran wonglakorn/Shutterstock
Printed and bound by CPI Group (UK) Ltd, Croydon, CR0 4YY

C9781119856627_040123

This book is dedicated to my wife Monika who has unfailingly provided me with practical assistance and encouragement in the preparation of this book.

I would like to convey my heartfelt thanks to my sons Paul, Simon, and Tristan in whom I have great pride and to my granddaughters Zena, Scarlet, Lente, and Heather who give me hope of a bright future.

Brief Contents

Preface *xv*
Acknowledgements *xix*

1 Clause 1 General Provisions *1*

2 Clause 2 The Employer *15*

3 Clause 3 The Engineer *19*

4 Clause 4 The Contractor *27*

5 Clause 5 Subcontracting *45*

6 Clause 6 Staff and Labour *49*

7 Clause 7 Plant, Materials, and Workmanship *53*

8 Clause 8 Commencement, Delays, and Suspension *59*

9 Clause 9 Test on Completion *69*

10 Clause 10 Employer's Taking Over *71*

11 Clause 11 Defects after Taking Over *75*

12 Clause 12 Measurement and Valuation *81*

13 Clause 13 Variations and Adjustments *87*

14 Clause 14 Price and Payment *97*

15 Clause 15 Termination by the Employer *109*

16 Clause 16 Suspension and Termination by the Contractor *115*

17 Clause 17 Care of the Works and Indemnities *119*

18 Clause 18 Exceptional Events *123*

19 Clause 19 Insurance *127*

20 Clause 20 Employer's and Contractor's Claims *133*

21 Clause 21 Disputes and Arbitration *139*

Appendix A. Guidance for the Preparation of Particular Conditions *145*

Appendix B. Employer's Claims *147*

Appendix C. Contractor's Claims *149*

Appendix D. Notices and Site Organisation *153*

Appendix E. Daywork and Contemporary Record Sheets *165*

Appendix F. Contractor's Costs *169*

Appendix G. Joint Ventures *173*

Model Form of a Pre-Bid Joint Venture Agreement *175*

Index of Clauses and Sub-Clauses *179*

Index of Sub-Clauses *183*

Contents

Preface *xv*
Acknowledgements *xix*

1   **Clause 1 General Provisions** *1*
1.1   Definitions *1*
1.1.10   The Contract *1*
1.1.12   Contract Data *1*
1.1.13   Contract Price *2*
1.1.15   Contractor's Documents *2*
1.1.20   Cost Plus Profit *2*
1.1.22   DAAB *2*
1.1.23   DAAB Agreement *2*
1.1.24   Date of Completion *2*
1.1.26   Daywork Schedule (If Included) *2*
1.1.27   Defects Notification Period (DNP) *3*
1.1.28   Delay Damages *3*
1.1.29   Disputes *3*
1.1.37   Exceptional Events *3*
1.1.38   Extension of Time (EOT) *3*
1.1.40   Final Payment Certificate (FPC) *3*
1.1.43   General Conditions *3*
1.1.45   Interim Payment Certificate (or IPC) *3*
1.1.46   Joint Venture *4*
1.1.48   Key Personnel *4*
1.1.49   Laws *4*
1.1.54   Month *4*
1.1.55   No objection *4*
1.1.56   Notice *4*
1.1.57   Notice of Dissatisfaction (or NOD) *4*
1.1.59   Particular Conditions *4*
1.1.62   Performance Certificate *5*

1.1.65   Plant   *5*
1.1.66   Programme   *5*
1.1.68   QM System   *5*
1.1.70   Review   *5*
1.1.71   Schedules   *5*
1.1.75   Special Provisions   *5*
1.1.77   Statement   *6*
1.1.78   Subcontractor   *6*
1.1.81   Tender   *6*
1.1.82   Tests After Completion   *6*
1.1.85   Unforeseeable   *6*
1.2   Interpretation   *6*
1.3   Notices and Other Communications   *6*
1.4   Law and Language   *7*
1.5   Priority of Documents   *8*
1.6   Contract Agreement   *8*
1.7   Assignment   *9*
1.8   Care and Supply of Documents   *9*
1.9   Delayed Drawings or Instructions   *9*
1.10   Employer's Use of Contractor's Documents   *11*
1.11   Contractor's Use of Employer's Documents   *11*
1.12   Confidentiality   *11*
1.13   Compliance with Laws   *12*
1.14   Joint and Several Liability   *12*
1.15   Limitation of Liability   *13*
1.16   Contract Termination   *13*

2   **Clause 2 The Employer**   *15*
2.1   Right of Access to the Site   *15*
2.2   Assistance   *16*
2.3   Employer's Personnel and Other Contractors   *17*
2.4   Employer's Financial Arrangements   *17*
2.5   Site Data and Items of Reference   *17*
2.6   Employer-Supplied Materials and Employer's Equipment   *18*

3   **Clause 3 The Engineer**   *19*
3.1   The Engineer   *19*
3.2   Engineer's Duties and Authority   *19*
3.3   The Engineer's Representative   *20*
3.4   Delegation by the Engineer   *20*
3.5   Engineer's Instructions   *21*
3.6   Replacement of the Engineer   *21*
3.7   Agreement or Determination   *21*
3.7.1   Consultation to Reach Agreement   *22*

3.7.2   Engineer's Determination   *22*
3.7.3   Time Limits   *22*
3.7.4   Effect of the Agreement or Determination   *24*
3.7.5   Dissatisfaction with Engineer's Decision   *24*
3.8   Meetings   *24*

4   **Clause 4 The Contractor**   *27*
4.1   Contractor's General Obligations   *27*
4.2   Performance Security   *27*
4.2.1   Contractor's Obligations   *27*
4.2.2   Claims under the Performance Security   *28*
4.2.3   Return of the Performance Security   *28*
4.3   Contractor's Representative   *29*
4.4   Contractor's Documents   *29*
4.4.1   Preparation and Review   *30*
4.4.2   As Built Records   *30*
4.4.3   Operation and Maintenance Manuals   *30*
4.5   Training   *31*
4.6   Co-operation   *31*
4.7   Setting Out   *32*
4.8   Health and Safety Obligations   *32*
4.9   Quality Management and Compliance Verification   *33*
4.9.1   Quality Management System – General Comments   *33*
4.9.2   Compliance Verification System   *34*
4.9.3   General Provisions   *34*
4.10   Use of Site Data   *34*
4.11   Sufficiency of the Accepted Contract Amount   *35*
4.12   Unforeseeable Physical Conditions   *35*
4.13   Rights of Way and Facilities   *38*
4.14   Avoidance of Interference   *38*
4.15   Access Route   *38*
4.16   Transport of Goods   *39*
4.17   Contractor's Equipment   *40*
4.18   Protection of the Environment   *40*
4.19   Temporary Utilities   *40*
4.20   Progress Reports   *41*
4.21   Security of the Site   *42*
4.22   Contractor's Operations on Site   *42*
4.23   Archaeological and Geological Findings   *43*

5   **Clause 5 Subcontracting**   *45*
5.1   Subcontractors   *45*
5.2   Nominated Subcontractors   *46*
5.2.1   Definition of a 'Nominated Subcontractor'   *46*

5.2.2   Objection to Nomination   *46*
5.2.3   Payments to Nominated Subcontractors   *46*
5.2.4   Evidence of Payment   *47*

6   **Clause 6 Staff and Labour**   *49*
6.1   Engagement of Staff and Labour   *49*
6.2   Rates of Wages and Conditions of Labour   *49*
6.3   Recruitment of Persons   *49*
6.4   Labour Laws   *50*
6.5   Working Hours   *50*
6.6   Facilities for Staff and Labour   *50*
6.7   Health and Safety for Personnel   *50*
6.8   Contractor's Superintendence   *51*
6.9   Contractor's Personnel   *51*
6.10   Contractor's Records   *52*
6.11   Disorderly Conduct   *52*
6.12   Key Personnel   *52*

7   **Clause 7 Plant, Materials, and Workmanship**   *53*
7.1   Manner of Execution   *53*
7.2   Samples   *53*
7.3   Inspections (by the Employer's Personnel)   *54*
7.4   Testing by the Contractor (on Site)   *55*
7.5   Defects and Rejection   *55*
7.6   Remedial Work   *56*
7.7   Ownership of Plant and Materials   *56*
7.8   Royalties   *57*

8   **Clause 8 Commencement, Delays, and Suspension**   *59*
8.1   Commencement of Works   *59*
8.2   Time for Completion   *59*
8.3   Programme   *60*
8.4   Advance Warning   *60*
8.5   Extension of Time   *61*
8.6   Delays Caused by Authorities   *62*
8.7   Rate of Progress   *62*
8.8   Delay Damages   *63*
8.9   Employer's Suspension   *64*
8.10   Consequences of Employer's Suspension   *64*
8.11   Payment for Plant and Materials after Employer's Suspension   *65*
8.12   Prolonged Suspension   *66*
8.13   Resumption of Work   *66*

9   **Clause 9 Test on Completion**   *69*
9.1   Contractor's Obligations   *69*
9.2   Delayed Tests   *70*

9.3   Re-testing   *70*
9.4   Failure to Pass Tests on Completion   *70*

10   Clause 10 Employer's Taking Over   *71*
10.1   Taking Over of the Works and Sections   *71*
10.2   Taking Over of Parts of the Works   *72*
10.3   Interference with Tests on Completion   *73*
10.4   Surfaces Requiring Reinstatement   *74*

11   Clause 11 Defects after Taking Over   *75*
11.1   Completion of Outstanding Work and Remedying Defects   *75*
11.2   Cost of Remedying Defects   *76*
11.3   Extension of Defects Notification Period   *77*
11.4   Failure to Remedy Defects   *77*
11.5   Remedying of Defective Work off Site   *77*
11.6   Further Tests after Remedying Defects   *78*
11.7   Right of Access after Taking-Over   *78*
11.8   Contractor to Search   *78*
11.9   Performance Certificate   *79*
11.10   Unfulfilled Obligations   *79*
11.11   Clearance of Site   *79*

12   Clause 12 Measurement and Valuation   *81*
12.1   Works To Be Measured   *81*
12.2   Method of Measurement   *83*
12.3   Valuation of the Works   *84*
12.4   Omissions   *85*
12.4.1 Contractor to give Notice of unreimbursed Costs   *85*

13   Clause 13 Variations and Adjustments   *87*
13.1   Right to Vary   *87*
13.2   Value Engineering   *89*
13.3   Variation Procedure   *90*
13.3.1 Variation by Instruction   *90*
13.3.2 Variation by Request for Proposal   *91*
13.4   Provisional Sums   *91*
13.5   Daywork   *92*
13.6   Adjustments for Changes in Laws   *93*
13.7   Adjustment for Changes in Cost   *94*

14   Clause 14 Price and Payment   *97*
14.1   The Contract Price   *97*
14.2   Advance Payment   *97*
14.2.1 Advance Payment Guarantee   *98*
14.2.2 Advance Payment Certificate   *98*
14.2.3 Repayment of the Advance Payment   *98*

14.3   Application for Interim Payment Certificates   *99*
14.4   Schedule of Payments   *100*
14.5   Plant and Materials intended for the Works   *101*
14.6   Issue of IPC   *102*
14.6.1   The IPC   *102*
14.6.2   Withholding (Amounts in) an IPC   *102*
14.6.3   Correction or Modification   *103*
14.7   Payment   *103*
14.8   Delayed Payment   *103*
14.9   Release of Retention Money   *104*
14.10   Statement on Completion   *105*
14.11   Final Statement   *105*
14.11.1   Draft Final Statement   *105*
14.11.2   Agreed Final Statement   *106*
14.12   Discharge   *106*
14.13   Issue of Final Payment Certificate (FPC)   *107*
14.14   Cessation of Employer's Liability   *107*
14.15   Currencies of Payment   *107*

15   Clause 15 Termination by the Employer   *109*
15.1   Notice to Correct   *109*
15.2   Termination for Contractor's Default   *110*
15.2.1   Notice   *110*
15.2.2   Termination   *111*
15.2.3   After Termination   *111*
15.2.4   Completion of the Works   *112*
15.3   Valuation after Termination for Contractor's Default   *112*
15.4   Payment after Termination for Contractor's Default   *113*
15.5   Termination for Employer's Convenience   *113*
15.6   Valuation after Termination for Employer's Convenience   *114*
15.7   Payment after Termination for Employer's Convenience   *114*

16   Clause 16 Suspension and Termination by the Contractor   *115*
16.1   Suspension by the Contractor   *115*
16.2   Termination by Contractor   *116*
16.2.1   Notice   *116*
16.2.2   Termination   *116*
16.3   Contractor's Obligations After Termination   *117*
16.4   Payment after Termination by the Contractor   *117*

17   Clause 17 Care of the Works and Indemnities   *119*
17.1   Responsibility for Care of Works   *119*
17.2   Liability for Care of the Works   *119*
17.3   Intellectual and Industrial Property Rights   *121*
17.4   Indemnities by Contractor   *121*

17.5   Indemnities by Employer   *122*
17.6   Shared Indemnities   *122*

18   Clause 18 Exceptional Events   *123*
18.1   Exceptional Events   *123*
18.2   Notice of an Exceptional Event   *124*
18.3   Duty to Minimise Delay   *124*
18.4   Consequences of an Exceptional Event   *124*
18.5   Optional Termination   *125*
18.6   Release from Performance Under the Law   *126*

19   Clause 19 Insurance   *127*
19.1   General Requirements   *127*
19.2   Insurance to be provided by the Contractor   *129*
19.2.1   The Works   *129*
19.2.2   Goods (Contractor's Equipment, Materials, Plant, and Temporary Works)   *130*
19.2.3   Liability for Breach of Professional Duty   *130*
19.2.4   Injury to Persons and Damage to Property   *130*
19.2.5   Injury to employees   *130*
19.2.6   Other Insurances Required by Law and by Local Practice   *131*

20   Clause 20 Employer's and Contractor's Claims   *133*
20.1   Claims   *134*
20.2   Claims for Payment and EOT   *134*
20.2.1   Notice to Claim   *135*
20.2.2   Engineer's Initial Response   *135*
20.2.3   Contemporary Records   *135*
20.2.4   Fully detailed claim   *135*
20.2.5   Agreement or determination of the Claim   *136*
20.2.6   Claims of Continuing Effect   *136*
20.2.7   General Requirements   *137*

21   Clause 21 Disputes and Arbitration   *139*
21.1   Constitution of the DAAB   *139*
21.2   Failure to Appoint DAAB Members   *140*
21.3   Avoidance of Disputes   *140*
21.4   Obtaining DAAB's Decision   *140*
21.4.1   Reference of a Dispute to the DAAB   *141*
21.4.2   The Parties' Obligations after the Reference   *141*
21.4.3   The DAAB Decision   *141*
21.4.4   Dissatisfaction with DAAB's Decision (NOD)   *142*
21.5   Amicable Settlement   *143*
21.6   Arbitration   *143*
21.7   Failure to Comply with DAAB's Decision   *144*
21.8   No DAAB in Place   *144*

Appendix A. Guidance for the Preparation of Particular Conditions *145*

Appendix B. Employer's Claims *147*

Appendix C. Contractor's Claims *149*

Appendix D. Notices and Site Organisation *153*

Appendix E. Daywork and Contemporary Record Sheets *165*

Appendix F. Contractor's Costs *169*

Appendix G. Joint Ventures *173*

Model Form of a Pre-Bid Joint Venture Agreement *175*

Index of Clauses and Sub-Clauses *179*

Index of Sub-Clauses *183*

Preface

The Conditions of Contract prepared by FIDIC are well established as the standard form of choice for use in the international construction industry.

Traditionally, in the standard FIDIC forms, the Engineer was given an authoritative role, enabling him to make informed judgements concerning the conduct and execution of projects with a large measure of independence from the Employer. From time to time, FIDIC updated these forms culminating in the 4th Edition 1987 (last reprinted 1992).

However, over the past 30–40 years, Employers became increasingly more involved in the day-by-day administration of projects, thereby restricting the powers of the Engineer to act independently of the parties.

In the same period, there was a significant increase in the availability of international funding, particularly for infrastructure projects. As a consequence, more and more companies undertook construction projects outside their national borders. However, many contractors were not always familiar with operating requirements of an FIDIC-based contract. Equally, Employers, well used to their national systems of contracting practices and law, faced with having to deal with contracts based on unfamiliar forms.

A key feature of the dispute resolution procedure contained in the FIDIC 4th Edition 1987, Sub-Clause 67.1 (Engineer's Decision) was the power and authority of the Engineer to make independent judgements. As the independence of the Engineer diminished following the increasing direct involvement of the Employer, the value of Engineer's decision was increasingly challenged by the contractors; as a result, more and more disputes were referred to arbitration.

Few in the construction industry regard arbitration as a satisfactory means of resolving disputes. Arbitration is a lengthy and expensive process that may or may not lead to awards. With a more flexible, realistic approach, the disputes could be negotiated more quickly and at a lower cost. A contractor also suffers because he is unable to foresee the outcome of the arbitration, and his cash flow is uncertain and damaged as a consequence of lengthy arbitration. Regrettably, there are instances of Employers preferring to refer some disputes to arbitration to avoid having to make decisions that they are unwilling to make themselves because of political or economic reasons.

Against this background, FIDIC undertook a major review of their standard forms.

FIDIC Conditions of Contract for Construction: First Edition 1999

Following extensive consultations with other parties engaged in the field of international construction, FIDIC issued, in 1999, a suite of four different contract forms:

CONS – Conditions of Contract for Construction (hereinafter referred to as 'The Red Book First Edition 1999'), which FIDIC recommends for use in building and engineering works designed by the Employer or his representative, the Engineer.

P & DB – Conditions of Contract for Plant and Design-Build ('The Yellow Book'), which FIDIC recommends for the provision of electrical and/or mechanical plant and for the design and execution of building or engineering works to be designed by the Contractor in accordance with the Employer's requirements.

EPCT – Conditions of Contract for EPC/Turnkey Projects ('The Silver Book'), which FIDIC recommends for the provision of a process or power plant on a turnkey project.

'Short Form of Contract' ('The Green Book') – A fourth contract form that intended for use in contracts involving small or repetitive work was also issued.

Subsequently, FIDIC has published other Conditions of Contract for specialist types of activity. The full listing of all other available contract forms is available on the FIDIC website, and these are not considered further in this book. Further, no consideration is given in this book to the second editions of the Yellow Book or Silver Book, which were also issued in 2017.

FIDIC Conditions of Contract for Construction: Second Edition 2017

In the 18 years that have elapsed since the publication of the First Edition 1999, inevitably, the substantial experience gained has shown that a further revision of this contract form was required.

In the Notes to this Second Edition 2017, FIDIC states that

'... *the edition provides:*

(1) *greater detail and clarity on the requirements for notices and other communications;*
(2) *provisions to address Employers' and Contractors' claims treated equally and separated from disputes;*
(3) *mechanisms for dispute avoidance and*
(4) *detailed provisions for quality management, and verification of Contractor's contractual compliance'.*

The principle features of this second edition include the following:

1. There is a greater emphasis on dispute avoidance.
2. There is a binding requirement for the Engineer to give an Engineer's Decision in respect of both Employer's and Contractor's claims within the prescribed time limits before referrals to the DAAB.
3. Clause 20 of the First Edition is now divided into two separate clauses. Clause 20 is now titled 'Employer's and Contractor's Claims', whilst new Clause 21 is titled 'Disputes and Arbitration', wherein the designation of DAB (Dispute Adjudication Board) has been amended to DAAB (Dispute Avoidance/Adjudication Board). This separation emphasises that Clause 20 concerns 'claims', which, not resolved after due procedure, become 'disputes' to be dealt within the procedures given in new Clause 21.
4. There are changed time limits for various events, e.g. submittal of claims and submittal of notices generally.
5. Programming requirements are more specific and will require more detailed and closer monitoring and reporting.
6. There is a broader definition of the word 'Notice', a consequence of which is the increase of workload for project management.
7. This Second Edition is a much bulkier document than its predecessor, having almost double the number of pages. The number of supplementary documents (including the 'Guidance for Preparations of Particular Conditions') has also increased.

Acknowledgements

The author is grateful to the Fédération Internationale des Ingénieurs Conseil (FIDIC) for permission to quote extracts from the Conditions of Contract for Construction, 'The Red Book' Second Edition 2017. Quoted extracts from this publication are given in italics.

Throughout this book, 'The FIDIC Conditions of Contract for Construction' ('The Red Book'), First Edition 1999, is referred to as the 'First Edition'

and

'The FIDIC Conditions of Contract for Construction' ('The Red Book'), Second Edition 2017, is referred to as the 'Second Edition'.

In this book, the Employer, the Engineer, the Contractor, and subcontractors are referred to in the masculine gender for simplicity and uniformity. The author wishes to emphasise that the book is intended to address ALL GENDERS on an equal basis as their male colleagues.

1

Clause 1 General Provisions

1.1 Definitions

Of the 58 definitions included in the First Edition, 34 have been included unchanged in the Second Edition and a further 18 included with minor changes only. Just one item, Sub-Paragraph 1.1.6.4 (Force Majeure), included in the First Edition, has been deleted from its original position and is now included in Clause 18 of the Second Edition under the revised heading 'Exceptional Events'.

There are 37 new or revised definitions included in the Second Edition and are listed below:

1.1.10 The Contract

The definition has been broadened to include two additional documents:

(a) 'The Contractor's proposal', which itself is not a defined term. It is assumed to be a general statement of how the Contractor intends to proceed with the execution of the Works. Conventionally, this is provided as part of the Contractor's tender.
(b) 'The JV Undertaking, if any'. A grouping of Contractors intending to tender as a Joint Venture will enter into a Pre-Bid Joint Venture Agreement to ensure that each member is committed to the Joint Venture. If the Employer requires a pre-bid qualification process, then the Pre-Bid Joint Venture Agreement can be used to support the JV qualification, otherwise it shall be included in the Joint Venture tender offer. The full JV Agreement will be finalised immediately after award should the tender be successful. An example of a Joint Venture Agreement is provided in Appendix G.

1.1.12 Contract Data

In the First Edition, 'Contract Data' was referred to as 'Appendix to Tender'. In the Second Edition the number of items included under this heading has been significantly increased (these are discussed later in Appendix A, 'Guidance for the Preparation of Particular Conditions').

Guide to the FIDIC Conditions of Contract for Construction: The Red Book 2017, First Edition.
Michael D. Robinson.
© 2023 John Wiley & Sons Ltd. Published 2023 by John Wiley & Sons Ltd.

1.1.13 Contract Price

This definition is minimal. A full definition is included in Sub-Clause 14.1 (The Contract Price).

1.1.15 Contractor's Documents

This definition has been enlarged seemingly to ensure every potential item has been included. Considering that the sub-clause ends *'and other documents of a technical nature'*, an extensive listing is hardly required.

1.1.20 Cost Plus Profit

This definition has been significantly improved from that given in the First Edition and provides clarity on an issue that has led to disputes in the past.

- The percentage for profit shall be as stated in the Contract Data.
- The percentage for profit shall be 5% if not otherwise stated in the Contract Data.

The Contractor should note that there have been instances where the Employer has failed to add any percentage for profit in the Contract Data. At a later date, the Employer could claim that because no percentage was given, the Contractor was not entitled to any payment under this item.

1.1.22 DAAB

'Dispute Avoidance/Adjudication Board' (DAAB) replaces the "Dispute Adjudication Board" described in the First Edition, as it more accurately defines the purpose of the Board as stated elsewhere in Clause 21 of the Second Edition.

1.1.23 DAAB Agreement

An updating of the DAB wording used in the First Edition.

1.1.24 Date of Completion

This is a new sub-paragraph not used in the First Edition and is self-explanatory.

1.1.26 Daywork Schedule (If Included)

This is a new definition not used in the First Edition but arguably should have been included. For further comment refer to commentary on Sub-Clause 13.5 (Daywork).

If a Daywork Schedule is not included in the Contract, it would be appropriate to query this omission in the Tender period.

1.1.27 Defects Notification Period (DNP)

The wording has been slightly modified but has the same intent as that stated in the First Edition.

1.1.28 Delay Damages

In the First Edition, the Delay Damages were specified in the (then) Sub-Clauses 8.2 and 8.7 and valued as stated in the Appendix to Tender. In the Second Edition, the relative clause numbers are Clauses 8.2 and 8.8 and any Delay Damages are to be valued as stated in the Contract Data (which is the retitled 'Appendix to Tender' – see Sub-Paragraph 1.1.12 above).

1.1.29 Disputes

This is a new definition not included in the First Edition. It is a very important topic and is to be referenced to Clause 21.

1.1.37 Exceptional Events

Clause 19 of the First Edition (headed 'Force Majeure') defined Force Majeure as an *exceptional event or circumstance*'. In this edition, the wording 'Force Majeure' has been retitled as 'Exceptional Events' with only very minor changes to the contents of the clause. The clause number is changed from Clause 19 to Clause 18.

1.1.38 Extension of Time (EOT)

This is a new definition. Sub-Clause 8.4 of the First Edition has been replaced by Sub-Clause 8.5 in the Second Edition. Refer to Sub-Clause 8.5 for further discussion. Note that the abbreviation EOT is now regularised.

1.1.40 Final Payment Certificate (FPC)

The wording is unchanged from the wording in the First Edition. Note that the abbreviation FPC is now regularised.

1.1.43 General Conditions

This is a new definition added presumably for the sake of completeness.

1.1.45 Interim Payment Certificate (or IPC)

Text is slightly modified with same intent as in the First Edition. Note that the abbreviation IPC is now regularised.

1.1.46   Joint Venture

This is a new definition. The First Edition did not contain this definition even though it is a frequently used term to describe a well-recognised Contractor's business structure. Note that the abbreviation JV is now regularised. It is observed that this Second Edition does not make reference to other Contractor's business structures such as a 'Consortium'.

1.1.48   Key Personnel

This is a new definition added presumably for completeness. Note that it specifically refers to the Contractor's personnel excepting only the Contractor's Representative.

1.1.49   Laws

This is a new definition added to reflect the changed heading of Sub-Clause 13.6 (Adjustment for Changes in Laws).

1.1.54   Month

This is a new definition. 'Month' is a word used occasionally in the Conditions of Contract.

1.1.55   No objection

This is a new definition and is a wording occasionally used in the Conditions of Contract.

1.1.56   Notice

This is a new definition. When read in conjunction with Sub-Clause 1.3 (Notices and Communications), this definition will have an important impact on the day-to-day management of the project. More than 100 instances where a Notice is required to be issued are identified in Appendix D of this book.

1.1.57   Notice of Dissatisfaction (or NOD)

This is a new definition and refers to the Notice one Party may give to the other Party if he is dissatisfied with either an Engineer's determination (Sub-Clause 3.7) or with a DAAB decision (Sub-Clause 21.4).

1.1.59   Particular Conditions

This is a new definition. No definition of 'Particular Conditions' was included in the First Edition, although the wording 'The Particular Conditions' was included in Sub-Clause 1.5 of the General Conditions. The same wording is included in the Second Edition, Sub-Clauses 1.5(a) and (b). This definition is solely a tidying up exercise.

1.1.62 Performance Certificate

This issue of the Performance Certificate is conditional on the Contractor having completed all his obligations, i.e. at the end of the expiry date of the last Defects Notification Period.

1.1.65 Plant

The wording of this definition has been slightly enlarged in the Second Edition. The definition refers to *'apparatus, equipment, machinery etc.'*, intended to form part or forming part of the permanent works.

1.1.66 Programme

This is a new definition. Although Sub-Clause 8.3 of the First Edition described the Contractor's general obligations in respect of programming, no specific definition of the word "Programme" was provided. A much-enlarged Sub-Clause 8.3 (Programme) is included in this Second Edition.

1.1.68 QM System

This is a new definition. The First Edition Sub-Clause 4.9, headed 'Quality Assurance', contained only a brief description of the Contractor's obligations. In this Second Edition, Sub-Clause 4.9, headed 'Quality Management and Compliance Verification Systems', contains a much more detailed description of the required QM system.

1.1.70 Review

This is a new definition. The word 'review' was variously used in the First Edition. However, a clear definition was not provided.

1.1.71 Schedules

The wording of this definition has been slightly amended.

1.1.75 Special Provisions

This is a new definition. The Second Edition has a number of annexures, including a document entitled 'Guidance for the Preparation of Particular Conditions'. Pages 13–52 of that document are subtitled 'Notes on the Preparation of Special Provisions', which is intended to provide guidance to the Employer and his advisors when preparing the Particular Conditions of Contract Part B for inclusion in the Contract Documents. There is no direct involvement of the Contractor in this matter.

1.1.77 Statement

This sub-clause has been expanded to make it clear that in addition to Statements for Applications for Interim Payments, the Contractor shall also provide Statements at Completion (Sub-Clause 14.10) and Statement at Final Completion (Sub-Clause 14.11).

1.1.78 Subcontractor

The term 'Subcontractor' is extended to include any designer appointed by the Contractor.

Note: Nominated Subcontractors are defined in Sub-Clause 5.2.1.

1.1.81 Tender

The definition contained in the First Edition has been extended to include:

• The Contractor's proposal
• The JV Undertaking (if applicable)

1.1.82 Tests After Completion

The definition has been rewritten with the same intent as in the First Edition.

1.1.85 Unforeseeable

The wording in the First Edition '. . .. *by the date for submission of the Tender'* is amended to '. . . *by the Base Date'*. Base Date is defined in Sub-Paragraph 1.1.4.

1.2 Interpretation

This sub-clause contains legal statements confirming (except where the context requires otherwise):

(a) words indicating one gender – includes all genders
(b) words indicating the singular also include the plural and vice versa
(c) "Agreements" have to be recorded in writing
(d) where something is stated to be <u>written</u> or <u>in writing</u>, this shall result in a <u>permanent record</u>

Two additional sub-clauses have been added in the Second Edition:

(e) *'may'* means that the Party or person has a choice of whether to act or not
(f) *'shall'* indicates an obligation to perform.

1.3 Notices and Other Communications

In the First Edition, this Sub-Clause was headed 'Communications'. In this Second Edition, the sub-clause has been significantly expanded and its application requires a much greater degree of formality between the Parties.

The opening paragraph contains an extensive listing of all potential types of communication and content which could arise during the course of executing a contract. Most items in the listing relate to routine matters arising during the performance of the contract and will be familiar to those directly engaged in its execution.

However, this Second Edition now introduces two related new items which require special attention, namely 'Notice' (refer Sub-Paragraph 1.1.56) and 'Notice of Dissatisfaction' (refer Sub-Paragraph 1.1.57).

A Notice is a written communication identified as a Notice which is sent by one Party to another.

Examples of Notices are:

Clause 8.1	The Engineer shall give a Notice to the Contractor stating the Commencement Date.
Clause 3.7.2	The Engineer shall give a "Notice of the Engineer's Determination" to both Parties of his determination of any claim.
Various Clauses	All claims submitted by the Contractor shall require a heading "Notice of Claim".

The general requirements of this sub-clause are:

- a Notice or communication must be in writing and a paper original signed by the Contractor's Representative (or authorised representative)
- an electronic original to be transmitted to/from the electronic addressees of the authorised representatives or
- delivered by hand against receipt,
- delivered, sent or transmitted to the address for the recipient's communications as stated in the Contract Data.

The management of 'Notices and Other Communications' is a burden on site staff and will require careful handling, especially if time related issues are involved. This subject is discussed further in Appendix D of this book.

1.4   Law and Language

The law governing the Contract shall be that of the country where the Contract is to be executed.

The wording of the First Edition has been upgraded. Both the ruling language and the language for communications are intended to be stated in the Contract Data. If not stated, then the language of these Conditions of Contract shall be the ruling language and the language for communications shall be the ruling language.

Should the stated governing law be not that of the country where the contract is executed, it may happen that the local courts may claim jurisdiction regardless of the wording of the Contract. Legal advice should be sought if such a situation arises.

1.5   Priority of Documents

Engineer or Contractor to give Notice of any discrepancy which may lead to Notice of Claim by Contractor.

The listing provided in the First Edition has been extended in the Second Edition as follows:

(a) the Contract Agreement   }

(b) the Letter of Acceptance   }   unchanged in the Second Edition

(c) the Letter of Tender   }

(d) The Particular Conditions item (d) of the First Edition has been amended. The First Edition Sub-Paragraph 1.1.1.9 describes a document titled 'Appendix to Tender', which is a listing of additional information pertinent to the Tender (and to be included in the signed Contract), and which are to be provided by the Employer.

In this Second Edition, this document is now re-titled 'Particular Conditions Part A – Contract Data'.

(e) The previous format of the Particular Conditions in the First Edition is retained and re-titled 'Particular Conditions Part B – Special Provisions'.

(f) the General Conditions   }

(g) the Specifications   }   Re-numbered but otherwise unchanged

(h) the Drawings   }

(i) The Schedules   }

(j) the JV Undertaking (if the Contractor is a JV) – New item (refer to Appendix G of this book)

(k) any other documents forming part of the contract – unchanged

'*If a Party finds an ambiguity or discrepancy on the documents, that Party shall promptly give a Notice to the Engineer, describing the ambiguity or discrepancy*'.

1.6   Contract Agreement

This Second Edition follows the same procedures given in the First Edition. The Employer is required to provide the Contractor with a Letter of Acceptance, which the Contractor should acknowledge in writing with the date of receipt duly noted. From the date of receipt of the Letter of Acceptance by the Contractor, a binding contract exists between the Parties. Within 35 days (was 28 days in the First Edition) after receipt of the Letter of Acceptance by the Contractor, the Parties shall sign a Contract Agreement based on the standard form annexed to the Particular Conditions of Contract.

It is preferable that the full contract documentation, including the Contract Agreement and those documents described in Sub-Clause 1.5, are all brought together in one comprehensive bound document. The various annexures need only be initialled by the signatories.

The signatories of both the Employer and the Contractor will have to be legally authorised to do so by their respective Board of Directors (or equal).

1.7 Assignment

The wording of the First Edition is unaltered.

'Neither Party is permitted to assign or transfer the whole or any part of the Contract or any benefit or interest in or under the Contract without the agreement of the other Party'.

Either Party may, as security, assign its right to any money due under the Contract to a bank or financial institution. Contractors generally will finance their activities with loans from their bankers. A typical condition of the provision of such loans is that the Contractor's income is to be channelled through these banks, thus providing a high degree of financial stability.

Government bodies may be reorganised and retitled from time to time, but this would not lead to reassignment.

However, if the Employer is not a government body, then events such as taking over by others, re-organisation, change of ownership etc. may occur. The Contractor is advised to obtain a legal review should these circumstances arise.

1.8 Care and Supply of Documents

Parties and Engineer are required to give Notice of errors or defects which may give rise to Notice of claim by Contractor.

In both the previous and Second Editions, the Specification and Drawings are to be in the custody of the Employer.

The Employer is required to provide the Contractor with two copies of the Contract and each subsequent drawing at no cost to the Contractor. Further copies may be provided if requested by the Contractor and at the Contractor's expense.

The Contractor is to retain the custody of the Contractor's Documents, unless and until taken over by the Employer.

In the Second Edition, the requirements of the First Edition have been partly amended. The Contractor is now required to provide the Employer with one paper original of each of the Contractor's Documents, together with

'one electronic copy (in the form stated in the Specifications or, if not so stated, in a form acceptable to the Engineer) and additional paper copies (if any) as stated in the Contract Data for each of the Contractor's Documents'.

One wonders if it would not be more efficient and less costly if both Parties were to exchange copies of drawings by electronic means, allowing each Party to decide the number of copies required at their own convenience.

1.9 Delayed Drawings or Instructions

The wording of this sub-clause in the Second Edition is broadly similar to the provisions stated in the First Edition.

Concerning the supply of drawings there are two possible scenarios:

1. The Employer will have appointed an Engineer responsible for both the design and supervision of the construction of the Works.

The Employer must ensure that:

(a) Drawings of sufficient quality and quantity are available for inclusion in the tender document package.

(b) Updated drawings are progressively issued for use as "drawings for construction". The first package of drawings will be required as soon as possible following the Contractor's receipt of the Letter of Acceptance.

Thereafter the Engineer and the Contractor should agree a schedule for the continuing supply of drawings which will reflect the Contractor's Programme of Works (which for this purpose can be assumed to approximate to the Programme of Works included in the Contractor's Tender offer). As a general indicator the Contractor will require drawings for construction at least two months prior to commencement of each segment of the Works in order that key materials (re-bar, embedments, etc.) can be ordered and delivered to Site. Due allowance should be made, where applicable, for shipping times and customs clearance. Drawing revisions are a common feature of any significant project and a late supply may delay Contractor activities. However, with good communication systems this possibility can be minimised. If the Contractor is delayed, he is entitled to give a Notice of Claim (all subject to Sub-Clause 20.2) to be given within 28 days.

2. The Employer may (for a number of reasons) decide to employ a design specialist to prepare the design of the Works and the Tender Documents (which implies that the Employer is satisfied with the design).

Thereafter, the Employer may engage the services of another engineer to supervise the construction stage of the project (the FIDIC Engineer of these Conditions of Contract). Any decision to separate the design of the Works from the supervision of construction frequently arises of separate financing packages for the two stages. However, occasionally, this separation can have unintended consequences.

The Author has experience of a modest road rehabilitation project in Eastern Europe which was designed by an external consultant using funding from an initial external financing package. After a further five years, the Employer was able to access a second financial package for the physical reconstruction of the road. A supervising engineer (the "Engineer") was duly appointed.

Soon after commencement of the work, the Engineer discovered that in the intervening five years the road had significantly deteriorated further, requiring significant changes to the design and the execution of additional quantities of work. The Employer had not taken this possibility into consideration when awarding the construction contract. The available funding package was inadequate to cover the resulting additional costs.

Further, the modified Red Book FIDIC 1999 contract document contained additional wording which inter alia stated that the Engineer could not order additional work without first obtaining the permission of the Employer. Due to a lack of funding the Employer was unable to provide the requested permission. The administration of the contract fell into disarray. After a prolonged delay, additional funding was eventually obtained to finish the contract which was delivered well over budget and very late.

Note:
Concerning restrictions on the Engineer's authority to order additional work, students of these FIDIC Contract Forms are invited to read the FIDIC publication 'THE FIDIC GOLDEN PRINCIPLES – First Edition 2019' which is available on the FIDIC website.

1.10   Employer's Use of Contractor's Documents

Although the Contractor retains the copyright and other intellectual property rights in the Contractor's Documents, the Employer has a free licence to use this information for the operation and maintenance of the relevant portion of the Works.

Item (c) refers to electronic or digital files, computer programmes, and other software which may be used at the on-site locations of the Employer and the Engineer.

Item (d) is a new item which describes the Employer's use of Contractor's Documents in the event of Termination for Contractor's Default (Sub-Clause 15.2) and Termination for Employer's Convenience (Sub-Clause 16.2) or Optional Termination (Sub-Clause 18.5).

In the First Edition there was a reference to '. . . *replacements of any computers supplied by the Contractor*', which obligation has been deleted from this Second Edition.

1.11   Contractor's Use of Employer's Documents

As in the First Edition, the Contractor is entitled to use Employer's Documents solely for the purpose of executing the Contract and for no other purpose without the written permission of the Employer.

1.12   Confidentiality

The corresponding sub-clause of the First Edition was titled 'Confidential Details' and consisted of one sentence.

In the Second Edition, this sub-clause has been considerably enlarged and re-titled 'Confidentiality', reflecting a more detailed, broader scope of the obligations of the Parties.

1st Paragraph

As in the First Edition, the Contractor is required to disclose all information (confidential or otherwise) to the Engineer, which is necessary to ensure Contractor's compliance.

2nd Paragraph

The Contract Documents may be used only for the execution of the Works. The Contractor is prohibited from publishing or disclosing '*any particulars of the Contract*'. The Contractor, including his head office, is advised not to give press or media interviews, but refer such issues to the Employer.

3rd Paragraph

Any information provided by the Contractor to the Employer and/or the Engineer and which is marked "Confidential" shall not be disclosed to third parties.

4th Paragraph

The above obligations do not apply if the confidential information is already in the public domain or readily available from other sources.

1.13 Compliance with Laws

The Parties have a general obligation to comply with *'all applicable laws'*.

(a) The Employer is required to obtain planning permits, zoning regulations, and similar regulations required for the execution of the Permanent Works together with any other permits etc. which are identified in the Specification. The Contractor has a general obligation to assist the Employer to obtain the required permits and other documentation including that which is identified in the Specification. The Contractor is required to comply with the permits etc. obtained by the Employer.

(b) The Contractor shall give all notices, pay all taxes, duties and fees required by Law in relation to the execution of the Works and shall indemnify the Employer in respect of a failure by the Contractor to comply with these requirements.

In preparing his tender, the Contractor should examine the Site in sufficient detail in order to determine if additional land is required for the siting of crushers, concrete plants, accommodation for personnel, etc., all of which may be subject to local planning regulations. The acquisition of land for quarrying purposes, borrow areas and waste areas which may be located at a distance from the Site, will also require the Contractor to obtain the appropriate permissions, licenses, etc. from the local authorities.

Royalties may be payable in respect of any excavated materials. This potential additional tax may be significant. The legal requirements are to be clarified and the Cost included in the Tender price. Sub-Clause 2.2 (Assistance), sub-paragraph b(i) states that *'If requested by the Contractor, the Employer shall promptly provide reasonable assistance to the Contractor to enable him to obtain permissions, licences etc'*.

'Should the Contractor suffer delay and/or incurs Cost as a result of the Employer's delay or failure to obtain any permit, permission etc. (see paragraph (a) above), the Contractor shall be entitled, subject to Sub-Clause 20.2, to EOT and/or payment of such Cost plus Profit'.

'If the Employer incurs additional costs as a result of the Contractor's failure to comply with any of his obligations given above, then the Employer shall be entitled, subject to Sub-Clause 20.2, to payment of these costs by the Contractor'.

1.14 Joint and Several Liability

(a) This sub-paragraph is a brief definition of a JV. If a grouping of contractors intends to submit a tender as a JV, then their intention will be made clear in a Pre-Bid Joint Venture Agreement submitted as part of the tender. This Pre-Bid Joint Venture Agreement will legally bind the contractor grouping and will be replaced by a full Joint Venture Agreement if the tender is successful.

(b) In complying with sub-paragraph (a), the authority of the JV leader will automatically be included in the JV Agreement.

(c) The members of the JV will be 'known' to each other prior to prequalification or at the latest prior to preparation of the tender. The JV will have its own legal identity at the date of submittal of tenders, usually with its own title such as (name of project) Joint Venture.

In a conventional JV, each of the JV members will provide a share of resources including Contractor's Equipment and senior staff to a pre-agreed plan. The Contractor's Representative (Sub-Paragraph 1.1.18) will almost certainly be provided by the lead company of the JV. The site organisation of administration, workshops, stores, technical support, and field crews will be a blend of the resources of the JV members. Exceptionally, it may happen that one member of the JV is to be engaged in a specialist activity. In the event of that specialist becoming bankrupt or otherwise unable to continue with the allotted work, then the JV would need to obtain consent of the Employer for the engagement of a replacement specialist company. To avoid such a possibility occurring, it is probably more manageable if the specialist were to be engaged as a subcontractor to the JV.

Should a member of the JV become bankrupt or otherwise unable to continue, the remaining members of the JV are obliged to share the share portion of the failed member between them. The share portion of the failed member is conventionally reduced to a nominal 0.01% of the total. This share portion becomes extinct only when the JV is wound up after completion of all activities, including settlement of all matters between the members of the JV.

Finally, if a grouping of contractors intends to tender for a project with the intention of independently constructing the total project in distinctly separate parts, this would constitute a Consortium and not a JV, requiring a different form of documentation.

1.15   Limitation of Liability

This is a new sub-clause, not included in the First Edition.

The opening sentence effectively states that neither Party is liable to the other Party for any loss of profit, consequential losses, etc. unless they are included in a listing of items provided in this sub-clause.

Key items are:

(a) Sub-Clause 8.8 (Delay Damages)
(b) Sub-section (c) of Sub-Paragraph 13.3.1 (Variation by Instruction)

1.16   Contract Termination

This is a legal statement. Subject to the requirements of the governing law, a termination of the Contract requires no other action than that given in the relevant sub-clause(s) of these Conditions of Contract.

Further details concerning Termination by Employer are provided in Clause 15 (Termination by the Employer) and details of Suspension and Termination by Contractor are provided in Clause 16 (Suspension and Termination by the Contractor). Optional Termination is the subject of Sub-Clause 18 (Optional Termination).

2

Clause 2 The Employer

2.1 Right of Access to the Site

This sub-clause refers not only to the Contractor's right of access to the Site but also his right to take possession of the Site.

The Employer may delay the hand-over of Site until the Contractor provides the Performance Security described in Sub-Clause 4.2. For larger projects, constructed by more experienced contractors, it is likely that one document accompanying the Tender will be a statement from a major insurance provider confirming that they will be providing the contract Performance Security, should their contractor client be awarded the project. The Employer will have less confidence in smaller contractors who may struggle to obtain the required Performance Security in the open market.

Understandably, the Employer may be reluctant to allow any contractor commence work without the necessary Performance Security (and insurances) in place.

The relevant date(s) for the Contractor taking possession of the Site is (are) to be provided in the Contract Data. If no date(s) are given in the Contract Data, then the Employer shall provide access to and possession of the Site in accordance with the preliminary programme of works submitted with the Contractor's tender. Attention is drawn to the wording of Sub-Clause 8.1, wherein it is stated that the Contractor shall commence the Works within 42 days after the Contractor receives the Letter of Acceptance. Thereafter, any delay in the provision of access to the Site or possession of the Site would entitle the Contractor to claim both time and costs by reference to this clause of the contract.

The handover of the Site is a significant event and should be properly managed by the Employer. It is the duty of the Employer to hand over the Site and not the Engineer's.

The Contractor should inspect the Site carefully and investigate any potential obstructions which have arisen after the date of any formal Site visit made prior to the date of the tender. Removal of these obstructions is not the Contractor's responsibility. Typically, an empty, unsecured site is a magnet for third parties who might use the site as an illegal dump site. Obstructions to a full site hand-over may arise as private properties may not have been purchased by the Employer and/or not yet vacated.

Guide to the FIDIC Conditions of Contract for Construction: The Red Book 2017, First Edition.
Michael D. Robinson.
© 2023 John Wiley & Sons Ltd. Published 2023 by John Wiley & Sons Ltd.

A formal signed protocol should be drawn up, identifying the time and date of the hand-over, with a statement of the condition of the Site. This protocol should be signed by authorised representatives of the Employer and the Contractor. For partial handovers, taking place on different dates, a separate protocol is required for each hand-over. Frequently, the Contractor may agree to accept the Site even though there exist obstructions which are the responsibility of the Employer. A common cause of obstruction arises from the lack of or incomplete land acquisition.

It is not always in the interest of the Contractor to commence work in a fragmented, inefficient manner. In severe cases, it may be appropriate for the Contractor to decline an incomplete hand-over.

In addition to handing over the Site, the Employer is also required to grant the Contractor the right of access to the Site. This subject is examined later under the heading of Sub-Clause 4.15 (Access Route).

2.2 Assistance

This sub-clause was titled 'Permits, Licences or Approvals' in the First Edition.

Frequently the assistance of the Employer is required to enable the Contractor to obtain the various permits, licences, and approvals necessary for the performance of the Contract.

The nature of the required permits, licences, and approvals will vary from country to country and from project to project and could include building permits, trade licences, licences for quarry operations, approvals from utility companies. These requirements should be researched using local knowledge and their potential value and impact on the timely performance of the Works evaluated in the preparation of the Tender.

It may be appropriate for the Contractor to raise any concerns during pre-tender meetings, so that the commitment of the Employer to assist in resolving problems is well established. Typical problems that frequently occur include:

- In many countries with a centralised planning, the supply of basic materials (cement, bitumen, fuels) may be strictly controlled and bulk supplies only available with the support of the Employer. Even then the authorities are often unwilling to pre-advise of any supply bottlenecks, which can be extremely frustrating.
- Most internationally financed projects are stated to be free of local taxes. Of particular interest are customs duties and value added tax (VAT). Often these arrangements cause problems between state ministries (e.g. the Treasury Ministry controlling the collection of taxes and revenues) and the Employer. Important supplies and equipment can be held up in part because the Treasury Department has failed to issue internal authorisation for duty-free imports. The Contractor (unless required by law) should not pay temporary deposits unless the Employer acknowledges liability to arrange for a refund. It is often very difficult to obtain refunds from the Treasury Departments. Again, during any pre-tender meeting the Employer could be asked to confirm that the appropriate arrangements are in hand. An unforeseen need to pay customs duties even on a temporary basis can affect the Contractor's cash-flow which can be damaging in the early stages of the Contract.

In many countries, utility companies may be tardy in dealing with Contractor's request for the provision of or relocation of services, possibly because of lack of materials or skilled workers. As a consequence, it would be expedient for the Contractor to assist in the resolution of these matters at the request of the Employer.

2.3   Employer's Personnel and Other Contractors

This sub-clause has been expanded from the single sentence given in the First Edition and re-titled.

The Employer is responsible for ensuring that the Employer's personnel and the Employer's other contractors cooperate with the Contractor (reference Sub-Clause 4.6) and comply with the same obligations which the Contractor is required to comply with under Sub-Clause 4.8 (Health and Safety) and Sub-Clause 4.18 (Protection of the Environment).

2.4   Employer's Financial Arrangements

Sub-clause 2.4 of the First Edition gave the Contractor the right to request the Employer to explain how payment for executing the Works was to be financed.

The revised and expanded sub-clause of the Second Edition confirms the Contractor rights given in the First Edition and additionally provides three specific examples of events which would entitle the Contractor to seek clarification.

If the Contractor:

(a) receives an instruction to execute a Variation with a price exceeding 10% of the Accepted Contract Amount, or the accumulated total of variations exceeds 30% of the Accepted Contract Amount
(b) does not receive payment in accordance with Sub-Clause 14.7 (Payment)
(c) becomes aware of a material change in the Employer's financial arrangements of which the Contractor has not received a Notice referencing this sub-clause,

then the Contractor may request the Employer to provide *'reasonable evidence'* that appropriate financial arrangements have been made. The Employer has 28 days to respond to the Contractor's request. Given the sensitivity of this subject, it would be appropriate if the Contractor were to discuss this issue with the Employer before making any written request.

2.5   Site Data and Items of Reference

In the First Edition, the provision of Site Data by the Employer was described in Sub-Clause 4.10. The Employer had a duty to make available to the Contractor all relevant data in the Employer's possession prior to the Base Date. The Employer was further required to make available to the Contractor additional data coming into the Employer's possession after the Base Date.

The Contractor was deemed to have inspected the Site and having taken the Employer's Data into account. The above requirements were supplemented by a listing of *'relevant matters'* to be assessed by the Contractor in preparation of his tender.

In the Second Edition, the equivalent clause is re-numbered and retitled as Sub-Clause 2.5 (Site Data and Items of Reference). The Employer's obligations have not been changed excepting that reference is also made to the *'original survey points, lines and levels of reference'* to be specified on the Drawings or supplied by a Notice from the Engineer.

The Employer is obliged to provide the Contractor with any additional data coming into his possession after Base Date.

Should the data provided by the Employer prove to be incorrect or incomplete, the Contractor would have grounds to claim any additional costs incurred and EOT in accordance with Sub-Clause 20.2. The list of topics given in the First Edition, now deleted, would provide useful guidance to the Contractor.

The Contractor's tender is based on the information available to him at the Base Date (28 days before the date of submittal of tenders). Tenderers may not be able to make allowance for any or all additional data supplied after Base Date. The possibility of a 'conditional' tender may arise in extreme circumstances.

2.6 Employer-Supplied Materials and Employer's Equipment

This sub-clause replaces the more explicit Sub-Clause 4.20 of the First Edition. The unit rates for the use of the Employer's Equipment will either be stated elsewhere in the Contract Documents or separately negotiated between Employer and the Contractor. Any insurance requirements will need clarification.

In the First Edition, the Employer-supplied materials were to be visually inspected on delivery for any shortages, default, or defect in the materials. Deficiencies were to be reported to the Engineer. The Employer would then immediately rectify the shortages, default, or defect. This procedure is not specifically stated in this Sub-Clause 2.6.

3

Clause 3 The Engineer

3.1 The Engineer

This sub-clause confirms the fundamental obligation of the Employer to appoint the Engineer to carry out the duties assigned to him. The Engineer may be a named person or may be a company. Should a company be named as the Engineer, then the company must advise the name of the person who will specifically be allocated the duties of the Engineer. It may be assumed that the Employer has already appointed the individual or company identified as the Engineer before tenders are submitted.

For larger contracts, the Engineer may be named in the Tender Documents which allows the Contractor the opportunity to assess any perceived risk arising from this appointment. Less satisfactory is the appointment of the Engineer in the post-tender period. The problem of a late appointment of the Engineer is exacerbated if the Engineer appointed for execution of the design is a different Engineer appointed for the supervision of the construction as valuable background information may be lost.

This separation of design and supervision, particularly for smaller projects, can lead to unsatisfactory situations.

3.2 Engineer's Duties and Authority

This sub-clause has been significantly changed from the equivalent sub-clause of the First Edition.

The First Edition stated in part *'If the Engineer is required to obtain the approval of the Employer before exercising a specific authority, the requirements shall be as stated in the Particular Conditions'*. Thus, the Employer was free to include restrictions in the Particular Conditions which prevented the Engineer from authorising any additional works (variations) or accepting Contract claims requiring additional payments to the Contractor.

The Second Edition strongly reinforces the authority of the Engineer.

Sub-Clause 3.2 states in part:

Third paragraph: *'If the Engineer is required to obtain the consent of the Employer before exercising a specified authority, it shall be stated in the Particular Conditions'.*

Guide to the FIDIC Conditions of Contract for Construction: The Red Book 2017, First Edition.
Michael D. Robinson.
© 2023 John Wiley & Sons Ltd. Published 2023 by John Wiley & Sons Ltd.

and

'There will be no requirement for the Engineer to obtain the Employer's consent before the Engineer exercises his/her authority under Sub-Clause 3.7 (Agreement or Determination)'.

Sub-Clause 3.7, among other issues, gives the Engineer the opportunity to make a fair determination of (any) matter or claim without the authority of the Employer.

Clause 3 of the 'Notes on the Preparation of Special Provisions' (Page 20) states that any instructions requiring the Engineer to first obtain the Employer's consent shall be clearly shown in the Special Provisions.

However, the above requirement is tempered by the final sentence of this sub-clause: *'It should be noted that any such requirement should not be applied to any action by the Engineer under Sub-clause 3.7, Agreement or Determination. . .'.*

Therefore, it is important that the Contractor's tender office carefully studies Clause 3 of the Particular Conditions of Contract.

3.3 The Engineer's Representative

This is a new sub-clause not included in the First Edition.

When read with Sub-Clause 3.1, it is now clear that the Engineer may or may not be resident at the Site. However, the Engineers Representative will be resident on the Site *'for the whole time that the Works are being executed at the Site'* and will be delegated with *'the authority necessary to act on the Engineer's behalf at the Site'*.

If the Engineer's Representative is temporarily absent from the Site during the execution of the Works, then the Engineer will appoint a suitable competent short-term replacement. The Contractor is to be given a Notice of such replacement.

3.4 Delegation by the Engineer

The Engineer may from time to time assign duties and delegate authority to assistants (and may revoke such authority) by giving Notice to the Parties (i.e. Employer and Contractor). It will be noted that there is no specific provision for the Engineer's Representative to delegate any of the duties and delegated authority assigned to him by the Engineer to these assistants.

The Engineer shall not delegate the authority to:

- act under Sub-Clause 3.7 (Agreement or Determination)
- issue a Notice to Correct under Sub-Clause 15.1 (Notice to Correct)

The assistants must be suitably qualified, competent, and are to be fluent in the language for communications. They are only authorised to issue instructions in respect of those matters assigned to them by the Engineer. (Reference may also be made to Appendix D of this book).

3.5   Engineer's Instructions

In the First Edition, this sub-clause was titled 'Instructions of the Engineer' and has been re-numbered.

The first sentence of this sub-clause states that *'the Engineer may issue the Contractor instructions for the execution of the Works'*.

The Engineer has the authority to delegate his authority (Sub-Clause 3.4). Consequently, the second sentence states that the Contractor shall only take instructions from:

- the Engineer
- or the Engineer's Representative
- or an assistant to whom authority has been granted by the Engineer to give instructions to the Contractor as described in Sub-Clause 3.4.

Should an instruction state that it constitutes a Variation, the Contractor shall follow the procedures given in Sub-Paragraph 13.3.1. If the Contractor considers that the instruction does constitute a Variation, then he shall promptly give Notice to the Engineer with his reasons before commencing any work described in the instruction. Within a period of seven days, the Engineer may confirm or revoke the instruction, in which case the Contractor is bound by the Engineer's response. The Contractor may then follow the claim procedures given in Clause 20. If the Engineer fails to respond, then the instruction is automatically revoked.

3.6   Replacement of the Engineer

If the Employer intends to replace the Engineer, he is required to give a Notice to the Contractor with details of his replacement. It is noted that the Engineer appoints both the Engineer's Representative (Sub-Clause 3.3) and assistants (Sub-Clause 3.4).

3.7   Agreement or Determination

Before using this clause, the attention of the reader is first drawn to the requirements of Sub-Clauses 20.1 and 20.2.

Summarised, Sub-Clause 20.2 (Claims for Payment and/or EOT) describes the procedures to be followed by the claiming Party in order to establish that he has a valid claim which shall be evaluated in accordance with Sub-Clause 3.7 (Agreement or Determination).

Within 84 days of the claiming Party becoming aware of the event or circumstance giving rise to the Claim, he shall submit a Notice of Claim to the Engineer (with reasons for the claim). Within a further 14 days, the Engineer is required to confirm the Contractor's Notice to be valid.

Within 84 days of the claiming Party becoming aware of the event or circumstance giving rise to the Claim, he shall submit a fully detailed claim to the Engineer or risk rejection of the claim. The Engineer may request further details. Should a dispute arise in respect of the validity of a claim, then the dispute shall be dealt with in accordance with the procedures contained in Sub-Paragraph 20.2.5 (Agreement or Determination of the Claim). Thereafter, the Engineer shall proceed in accordance with the provisions of Sub-Clause 3.7 (Agreement or Determination).

3.7.1 Consultation to Reach Agreement

The Engineer has a duty to *'consult with both Parties'* and *'encourage discussion between the Parties'*. Unless otherwise agreed, the Engineer shall keep a written record of any consultation.

If an agreement can be reached between the Parties (subject to time limitations of Sub-Paragraph 3.7.3), the Engineer shall prepare a 'Notice of the Parties' Agreement' which shall be signed by both Parties.

If no agreement is reached, then the Engineer shall confirm this lack of agreement and proceed to make *'a fair determination of the claim or matter'* within the stated time limits.

3.7.2 Engineer's Determination

There are three possible outcomes resulting from an Engineer's Determination:

(i) The Engineer's Determination is acceptable to the Parties. A written agreement shall be prepared by the Engineer and signed by the Parties.

(ii) Should either Party be dissatisfied with a determination of the Engineer, they shall give a 'Notice of Dissatisfaction with the Engineer's Determination (NOD)', setting out reasons for dissatisfaction. Time limits apply. Thereafter either Party may proceed to obtain a DAAB decision (Sub-Clause 21.4 refers).

(iii) An Engineer's Determination may only be partly acceptable to both Parties. However, there may be elements of the acceptable portion, which cannot be separated from the non-acceptable portion. In such case, these elements shall be treated as part of the non-acceptable portion. The acceptable portion will be reduced correspondingly and shall become final and binding on the Parties as if the NOD had not been given. The disputed non-acceptable portion may be referred to the DAAB (Sub-Clause 21.4 refers).

3.7.3 Time Limits

Note:
It is difficult to absorb the contents of this Sub-Clause 3.7 in a first reading. In particular, which time limits (Sub-Paragraph 3.7.3) are applicable in which situation can be confusing and may lead to unintentional errors.

It is recommended that the reader first reads through the whole, ignoring references to time limits in the first instance, in order to obtain a good understanding of the logic underlying the whole sub-clause. Reference may be made to the following logic diagram which culminates either in a settlement of claim or referral to DAAB. The various time limits (Sub-Paragraph 3.7.3 refers) may be added to the convenience of the reader.

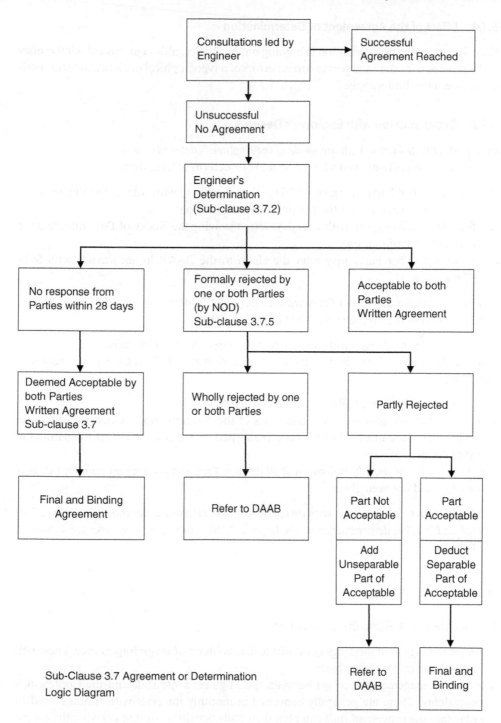

Sub-Clause 3.7 Agreement or Determination
Logic Diagram

3.7.4   Effect of the Agreement or Determination

Each agreement or determination is binding on the Parties, although this sub-clause does make provision for the Engineer to correct errors of a typographical or clerical or arithmetical nature. Time limits apply.

3.7.5   Dissatisfaction with Engineer's Decision

First part: *Dissatisfaction with the whole of the Engineer's determination*
   If either Party is dissatisfied with the whole of a determination, then

(a) the dissatisfied Party may give a NOD to the other Party with copy to the Engineer
(b) this NOD. . .. shall set out the reasons for dissatisfaction
(c) the NOD shall be given within 28 days after receiving the Notice of Determination (or corrected determination)
(d) thereafter either Party may refer the claim to the DAAB in accordance with Sub-Clause 21.4

   If no NOD is given by either Party within a period of 28 days stated in (c) above, then the determination of the Engineer shall be binding on them.

Second part: *Dissatisfaction with only parts of the Engineer's determination*
   If the dissatisfied Party is dissatisfied with only parts of the Engineer's determination, then

- this unacceptable part shall be identified in the NOD
- this unacceptable part and any other parts of the determination (which are otherwise acceptable but are affected by the unacceptable part) are deemed to be separable from the rest of the determination
- the remainder of the determination shall become final and binding on both Parties as if the NOD had not been given.

   'Should a Party fail to comply with an agreement of the Parties under this Sub-Clause 3.7 or a final and binding determination of the Engineer', the other Party may refer the failure to the DAAB.

3.8   Meetings

Meetings tend to fall into three categories:

- Senior management meetings convened to discuss items of major importance. These will be convened on an ad-hoc basis.
- Formal management meetings between the Engineer's site staff and the Contractor's equivalents. These are generally convened at monthly intervals with timings fixed in order that the concerned staff can plan their daily activities without delaying the execution of the Works.

Considering that this sub-clause requires the Engineer to *'keep a record of each management meeting. . .'*, either the Engineer or more likely the Engineer's Representative should attend and act as chairman of the meeting. The Contractor's Representative's presence would also be appropriate.

As work on the project expands, there is a tendency for too many people to attend these meetings which can greatly extend their duration and occasionally disrupt the decision-making process. Consideration could be given to dividing Site Meetings into two separate segments:

(a) Technical Meetings. Key topics would include progress of the Works, programme requirements, availability of plant, access problems, material sources and availability, Drawings etc.

(b) Administration Meetings. These would centre on contractual issues, certification preparation etc. These matters have little interest to the technical staff, who can be better employed elsewhere. Where useful, a member of the administrative personnel could attend the technical meetings in order to provide continuity.

Subcontractors, representatives of utility companies, etc. could be invited to either meeting as required.

The chairman of the meeting may prepare the minutes of the meeting himself or hand the task to a competent assistant. A broad agreed statement of the proceedings of the meeting should be prepared as the meeting proceeds. This would allow key decisions to be actioned without delay. The final Minutes of Meeting can then be prepared and distributed without delay or complaint of their content from any party.

Irregular meetings with service providers, local or central government agencies and similar bodies are frequently necessary to resolve obstructions to the progress of the Works. Powerlines, water supplies are required or need relocating. Land acquisition may also have to be considered. A formal minutes of meeting should be drawn up as a matter of record if only to ensure each of these external agencies adhere to agreements made. This is important, particularly if additional costs are involved.

4

Clause 4 The Contractor

4.1 Contractor's General Obligations

This sub-clause provides an extensive description of the Contractor's duties and responsibilities during the execution of the Works. These obligations are broadly the same as those given in the First Edition but are now defined in far greater detail. From the Contractor's viewpoint there are no major changes in this Second Edition that would adversely affect his operations.

This sub-clause confirms that, unless specifically stated elsewhere, the Contractor does not have any responsibility for the design or specification of the Permanent Works. Exceptionally, if the Contract Documents do require the Contractor to design any part of the Permanent Works, the Contractor is required to prepare a detailed listing of the Contractor's Documents, time for the execution of this requirement (including testing, training, as built drawings, etc.), and time for execution which must be provided for in the Programme of Works.

4.2 Performance Security

The Contractor shall provide the Employer with a Performance Security in the amount and currencies stated in the Contract Data. If no amount is stated in the Contract Data, this sub-clause shall not apply.

4.2.1 Contractor's Obligations

The Contractor shall deliver the Performance Security to the Employer (copied to the Engineer) within 28 days after receiving the Letter of Acceptance. The Performance Security shall be issued by an entity and from within a country to which the Employer shall give consent. The Contractor may be in a position to name the entity who will be providing the Security in his Tender submittal.

The Performance Security shall remain valid and enforceable until the issue of the Performance Certificate and the Site has been finally cleared in accordance with

Guide to the FIDIC Conditions of Contract for Construction: The Red Book 2017, First Edition. Michael D. Robinson.
© 2023 John Wiley & Sons Ltd. Published 2023 by John Wiley & Sons Ltd.

Sub-Clause 11.11. It is to be noted that the requirement for the Site to be cleared as a prequel to the return of the Performance Security was not included in the First Edition. Site clearance can be a tedious operation and should not be allowed to cause an extension to the validity of the Security.

A further requirement not included in the First Edition relates to Variations and/or adjustments under Clause 13 (Variations). Should the total amount of Variations and/or adjustments increase/decrease the Contract Price more than 20% of the Accepted Contract Amount, then <u>at the Employer's request</u> (underscore added) the amount of Performance Security shall be increased/decreased by the percentage of the accumulative increase.

If the Contractor incurs additional cost as a consequence of any increase in the Accepted Contract Amount, then Sub-Paragraph 13.3.1 (Variation by Instruction) shall apply as if the increase had been instructed by the Engineer. Should the amount of the Variations and/or adjustments decrease the Accepted Contract Amount, then the Contractor may, with the agreement of the Employer, decrease the amount of the Performance Certificate by the same percentage.

4.2.2   Claims under the Performance Security

The Employer is entitled to make claim under the Performance Security in five different circumstances:

(a) failure by the Contractor to extend the validity of the Performance Security (Reference: Sub-Paragraph 4.2.1, second section)
(b) failure by the Contractor to pay the Employer any amount as agreed or determined under Sub-Clause 3.7 or agreed or decided under Clause 21 (Disputes and Arbitration)
(c) failure by the Contractor to remedy a default stated in a Notice under Sub-Clause 15.1 (Notice to Correct)
(d) circumstances which entitle the Employer to terminate the Contract under Sub-Clause 15.2 (Termination for Contractor's Default)
(e) failure by the Contractor to comply with Sub-Clause 11.5 (Remedying Work off Site).

Annexures C and D (Pages 59 and 60) of the Particular Conditions refer.

4.2.3   Return of the Performance Security

The Employer shall return the Performance Security within 21 days of the Contractor's compliance with Sub-Clause 11.11 (Clearance of Site)
   or
promptly if the Contract is terminated in accordance with

- Sub-Clause 15.5 (Termination for Employer's Convenience)
- Sub-Clause 16.2 (Termination by Contractor)
- Sub-Clause 18.5 (Optional Termination)
- Sub-Clause 18.6 (Release from Performance under the Law)

4.3 Contractor's Representative

The requirements stated in the First Edition have been retained but have been extensively re-stated in a more explicit manner in the Second Edition.

The Contractor is required to appoint the Contractor's Representative and provide him with the necessary authority to act on behalf of the Contractor.

In addition to having a necessary qualification and appropriate experience, the Contractor' Representative must also be fluent in the *'language for communications'* and *'the whole of his time shall be given to directing the Contractor's performance of the Contract'*.

The Contractor's Representative will have his own letter of employment provided by the Contractor. This is a private document and is not subject to review or inspection by any other party.

This letter of employment will necessarily place restrictions on the internal powers and authority of the Contractor's Representative. In essence, the Contractor's Representative will not be empowered to take actions which might endanger the economic interests or financial stability of the Contractor. He will be authorised to operate a local bank account, purchase local goods and services, and to pay local staff and any other locally based matters. All offshore interests will be safeguarded by the Contractor's Head Office. The apparent gap between the Contractor's Representative's total authority described in this sub-clause and the restricted authority described in his own personal letter of employment will be managed by the Contractor's Head Office. Therefore, the standard practice is that the Contractor's Representative will immediately refer any matter arising beyond his authority to the Contractor for their further instruction. In extreme or delicate topics, the Employer may prefer to discuss these matters directly with the Contractor's Head Office senior management.

The Contractor shall not, without the prior consent of the Engineer, revoke the appointment of the Contractor's Representative. Reasons for the departure of the Contractor are typically: end of personal contract, ill health, or personal reasons – as in other professions. The departure of the Contractor's Representative will generally be a planned event of which the Employer and the Engineer will be informed and a proposed replacement for whom approval has already been requested by the Contractor will be appointed.

The Contractor's Representative will inevitably be away for short periods for a number of reasons, including off-site meetings, medical reasons, short holidays, etc. In such instances, the Contractor's Representative is required to delegate his authority to a suitably qualified senior member of his staff to provide short term cover to act in his absence.

4.4 Contractor's Documents

A broad definition of "Contractor's Documents" was provided in the First Edition (Sub-Paragraph 1.1.6.1) and expanded in the Second Edition (Sub-Paragraph 1.1.15) and this Sub-Clause 4.4 added.

4.4.1 Preparation and Review

Four categories of Contractor's Documents are to be provided for review by the Engineer:

(a) Those documents stated in the Specification. The Specification is a tender document which is to be included in the final Contract Document. Each specification is compiled to define the requirements of individual projects. The Contractor's tender office will routinely review the contents of the given Specification in the preparation of tenders.

(b) Those documents which are obtained by the Contractor in compliance with local laws. These requirements should be clarified by the Contractor prior to submission of tenders. The Employer is likely to be able to provide informal guidance.

(c) As-built Records – see 4.4.2 below.

(d) Operation and Maintenance Manuals – see 4.4.3 below.

The Contractor's Documents are to be written in the language for communications (see Sub-Clause 1.4). When buying materials on the open market for inclusion in the Permanent Works, the Contractor should ensure that supporting documentation is provided in the appropriate language.

4.4.2 As Built Records

Unless stated in the Specification, the Contractor is not responsible for the preparation of as-built records. (However, it would be quite unusual if this task was not allocated to the Contractor.)

Otherwise, 'the Contractor shall prepare, and keep up to date a complete set of "as built" records of the execution of the Works, showing the exact as-built locations etc. as executed by the Contractor' and 'details of the as-built records shall be as stated in the Specification (if not stated, as acceptable to the Engineer')

There are, at any one time, a significant number of work activities which could fall under this heading including, for example, concrete pour preparations and installation work, concreting operations, survey works, laboratory and material testing etc. It would be appropriate that the Engineer and Contractor agree a format for the efficient maintenance of these records in a standardised manner, at the outset of the project.

A second topic which requires discussion under this heading is the issue of "as-built" drawings. Considering that the Contractor has no authority to amend the design of the Works, it is incumbent on the Engineer to instruct any required modifications to the design whether by variation or otherwise. Consequently, the Engineer has the duty to modify the design drawings and design calculations. The Contractor is required to prepare "as-built drawings" and provide related data including survey and measurement records together with design drawings and data relating to any part of the Works for which design responsibility was allocated to him under the Contract.

4.4.3 Operation and Maintenance Manuals

Unless stated in the Specification, the Contractor is not required to provide any Operation and Maintenance Manuals.

Operation and Maintenance Manuals are routinely required in respect of turbines, pumps, gates, and other items of machinery to be provided and installed as part of the contract. This work will routinely be undertaken by specialist subcontractors generally (but not always) acting as nominated subcontractors under the Contract. Nonetheless, there may be smaller items of machinery to be installed by 'in-house' subcontractors. In both instances, the operation and maintenance manuals will be prepared by the subcontractors (supervised by the Contractor) for the approval with the Engineer with the involvement of the Employer who is the end user.

4.5 Training

Unless stated in the Specification, the Contractor is not required to provide any training of the Employer's personnel.

The extent of the training of the Employer's staff is to be stated in the Specification. The educational quality and previous work experience of the Employer's staff selected for training is crucial. Those who are to be trained to maintain and operate complex machinery should be selected at an early stage, so that they may work alongside the Contractor's employees during the installation stage.

Those suitable for routine maintenance duties will need less training. The Contractor may be willing to transfer some of his own employees, who have already suitable experience, directly to the Employer for employment and additional training as the date of the Taking Over of the Works draws closer.

4.6 Co-operation

The general obligation of the Contractor to provide 'Appropriate opportunities for other parties to carry out work on or near the Site' has a number of aspects:

1. Ideally, any significant requirement to provide services to other parties on the Site should be described in the Tender Documents in order that the Contractor may include the cost thereof in the tender offer.
2. Access roads constructed on Site for the Contractor's own use may be used by others for general access purposes, but other roads specifically required for use by other parties are to be constructed and maintained by them.
3. Facilities provided on Site by the Contractor for the use of his own employees are not for use by others unless agreed by the Contractor or specified in the Contract.
4. Accesses on Site, Site security, etc. provided on Site by the Contractor for Health and Safety purposes are for use by all parties. This does not negate the responsibility of the other parties to safeguard their own staff.
5. The Contractor is not obligated to allow other parties use of his Equipment and Goods.

The above are indicative of issues which regularly occur on projects where more than one contractor is engaged on the same Site, and which may lead to an adverse situation where the Contractor may suffer delay or additional costs and entitled to claim by giving a Notice to the Engineer in accordance with Sub-Clause 20.2.

The Contractor shall, as stated in the Specification or as instructed by the Engineer, provide 'appropriate opportunities' for other parties to carry out work on or near the Site.

This sub-clause continues by stating that appropriate opportunities may include the use of Contractor's Equipment. . . .. and/or other Contractor's facilities or services on the Site.

It is appropriate that detailed requirements (if any) are given in the Specifications in order that they may be evaluated by the Contractor and the cost thereof included in the Tender price.

Contractors are generally unwilling to allow their key equipment to be used by others unless supervised and operated by their own staff, not least because of the insurance complications which may otherwise arise. It is inevitable that the Contractor will give his own needs precedence. Additional costs for relocation and return of the equipment will have to be met by the Employer.

4.7 Setting Out

Sub-Clause 2.5 requires that the Employer makes available to the Contractor the original survey control points. The Employer should arrange for a check to be made before handover to ensure these control points physically exist and have not been disturbed by third parties.

One of the first operations undertaken by the Contractor on the Site will be check and confirm the survey control points which will enable site clearance to commence. Preferably this is an operation which should be jointly undertaken by the Engineer's and Contractor's surveyors, so that corrective measures can be quickly determined. Should the Contractor find there is an error (or errors), he should give the appropriate Notice to conform with the requirements of Sub-Clause 3.7 and discuss with the Engineer the corrective actions. The bureaucratic process stated in this sub-clause could and should be quickly resolved both to minimise any delays and to avoid the cost of equipment standing idle and subsequent loss of time.

4.8 Health and Safety Obligations

The First Edition (Sub-Clause 4.8) provided a relatively brief listing of the obligations of the Contractor under the heading "Safety Procedures". This was an update of the wording of the early FIDIC document last reprinted in 1992. A review of Sub-Clause 19.1 of this early document is concerned with safety and security but the subject of health is not mentioned. Times have changed. All those engaged in the construction industry and the wider public beyond now expect and demand high standards in respect of Health and Safety. This Sub-Clause 4.8 provides a much wider view of the Contractor's obligations under the heading of Health and Safety. The contents of the Health and Safety manual will apply to all those engaged on the project (or visiting the Site), including not only the Contractor's own personnel but also those of his subcontractors and suppliers and the Engineer's and Employer's personnel on Site.

Contractors, when working in their own countries of origin, are likely to have available a standardized Health and Safety manual which can be modified according to the needs of individual projects. However, an international contract, using the FIDIC Conditions of Contract, is likely to be executed in a poorer, less wealthy country where health services, social services, etc. are not readily available and consequently a different type of Health and Safety manual will require preparation.

Depending on the size of the project, it may be both convenient and economical for a specialist company with experience of the country of execution to be tasked with the preparation of such a manual.

4.9 Quality Management and Compliance Verification

4.9.1 Quality Management System – General Comments

General Comments

The corresponding clause in the First Edition consisted of nine lines of print under the heading of 'Quality Assurance'. In this Second Edition, the Contractor's obligations in respect of quality control and verification have been significantly expanded, reflecting the increasing importance of these topics in today's construction industry.

The FIDIC Conditions of Contract are designed for use on projects in countries where local Conditions of Contract, including quality management provisions, are likely to be inadequate for large, sophisticated, and internationally financed projects. Further, the international loan agencies overwhelmingly support the use of the FIDIC Conditions of Contract.

Larger contractors will have significant experience of QMS/CVS, together with knowledgeable staff able to modify standard documentation as required for individual projects.

Smaller contractors, who may be locally based, can be expected to have difficulty to comply with the specified QMS/CVS requirements. The assistance of an external consultant may be required to train local staff. Both the Engineer and the Contractor will require dedicated staff in order to manage the QMS/CVS system at the performance level specified in these Sub-Paragraphs 4.9.1 and 4.9.2.

Detailed Requirements

The preparation of the required documentation will be on hold until the Contractor receives a Letter of Acceptance.

The Contractor is to provide the Engineer with the QMS document within 28 days of the Commencement Date. Sub-Clause 8.1 (Commencement of Works) states that *'the Commencement Date shall be within 28 days after the Contractor receives the Letter of Acceptance'* and *the Commencement Date shall be not less than 14 days before the Commencement Date'*.

Therefore, the Contractor has 14 days plus 28 days = 42 days to present his QMS documents to the Engineer.

Considering that a large amount of data will not be available at the Commencement Date, the QMS document will require continuous updating as additional data becomes available.

The QMS shall be in accordance with the details in the Specification (included in the Tender Documents).

Contractor's procedures are to include the following:

- all notices and other communications to be traceable
- proper coordination and interfaces between stages of work
- submittal of documents to the Engineer for review
- internal audits of QMS regularly (at least once every six months). Contractor to provide report to Engineer containing measures to rectify and improve the QMS
- provide for external audit if required

Engineer's procedures are to include the following:

- Review of the QMS documents submitted by the Contractor
- Notice to the Contractor concerning faults in the QM system
- Notice to the Contractor in the event of non-compliance (for immediate correction by the Contractor)

4.9.2   Compliance Verification System

This sub-paragraph refers to the quality control (and associated work and workmanship) of Materials either produced on Site by the Contractor or provided by external suppliers delivered to the Site which are intended to be incorporated in the Works. The Contractor shall carry out inspections and testing to verify that these materials comply with the Specifications.

The Contractor shall provide the Engineer with a complete set of compliance verification documentation for the Works or Section in the manner described in the Specifications or as agreed with the Engineer.

4.9.3   General Provisions

This is a general reminder to the Contractor that compliance with Sub-Paragraph 4.9.1 (QMS) and Sub-Paragraph 4.9.2 does not relieve the Contractor from any of his duties and responsibilities under the Contract.

4.10   Use of Site Data

Note:
The obligation of the Employer to supply Site Data is described in Sub-Clause 2.5 of the Conditions of Contract. The term Site Data also includes basic survey data. Sub-Clause 4.7 (Setting Out) describes the verification of this survey data.

The Contractor is solely responsible for the analysis of the Data provided by the Employer and the Data which he has himself obtained from other sources, including such as government agencies.

It is a truism that a tenderer given unlimited time and unlimited resources could discover everything necessary for a risk-free project. This is clearly not a practical consideration for a tenderer preparing his offer in a limited period of time. FIDIC recognises this reality by stating that '. . . *to the extent which was practicable (taking into account of cost and time), the Contractor shall be deemed to have obtained all necessary information. . .*'.

This criterion of practicality has a profound influence on any claim the Contractor may wish to make under the provisions of Sub-Clause 4.12 (Unforeseeable Physical Conditions). The Contractor's estimating office would be well advised to keep copies or records of the data provided by the Employer together with copies of records of data obtained elsewhere that had an influence on his tender. These may be important in the evaluation of claims.

Under this sub-clause the 'Contractor is deemed to have inspected and examined the site before submitting his Tender offer'.

It is not an obligatory duty, but the majority of employers organise a formal inspection of the site which the Contractor may be obliged to attend as a precondition of the tendering process. Should the Site be in or adjacent to a restricted area, the formal inspection may be the only opportunity for the Contractor to inspect the site. It is recommended that the Contractor properly prepare himself for any site inspection. Clarifications should be sought where appropriate from the Employer/Engineer. A written record should be prepared complete with a photographic record for future reference.

4.11   Sufficiency of the Accepted Contract Amount

The Second Edition uses the same wording as the First Edition. Accuracy and completeness are skills of the Contractor's Tender Office.

4.12   Unforeseeable Physical Conditions

If the Contractor encounters adverse physical conditions which he considers to be 'unforeseeable', he shall give a Notice to the Engineer, which shall

- be given promptly, so that the Engineer may inspect the unforeseeable conditions
- describe the unforeseeable conditions and why the Contractor considers them unforeseeable
- evaluate any delays arising from the unforeseeable conditions and any increase in cost.

The Contractor shall be entitled to payment of additional costs and EOT referenced to Sub-Clause 20.2.

'Unforeseeable' is defined in Sub-Paragraph 1.1.85 as '. . . *not reasonably foreseeable by an experienced contractor*' by the Base Date. It will be noted that definition refers to hypothetical experienced contractor and not the Contractor himself. In the presentation of any claim under this heading, the Contractor's logic should be to demonstrate that a typically experienced contractor could not have foreseen the unforeseeable condition and therefore he, the Contractor, also an experienced contractor, would equally not have foreseen the unforeseeable conditions. What the Contractor himself may or may not have foreseen is not of immediate concern.

Frequently secondary disputes may arise over the practical application of the word 'reasonably' contained in Sub-Paragraph 1.1.85. In attempting to provide guidance on this point, the latest FIDIC Guide expressed the opinion that for a contract of three years' duration an experienced contractor might be expected to foresee an event which occurs on average once every six years. An event which occurs once every ten years might be regarded as 'unforeseeable'. Another authority has commented that the reference is to what was reasonably foreseeable by an experienced contractor and not by a research professor at university.

Secondly, the cut-off date for foreseeability is the date of tender and not the Base Date. This criterion appears to be harsh on the Contractor since it implies he has to conduct one last site inspection just before submitting his tender offer, which is clearly impractical.

FIDIC defines physical conditions as 'natural physical conditions and man-made or other physical conditions and pollutants which the Contractor encounters at the Site, including sub-surface and hydrological conditions, but excluding climatic conditions.

"Sub-surface conditions" are those conditions below the surface, including those with a body of water and those below the riverbed or seabed.

"Hydrological conditions" means the flows of water, including those which are attributable to off-Site climatic conditions.

"Physical conditions" excludes climatic conditions at the Site and therefore excludes the hydrological consequences of climatic conditions at Site'.

The foregoing leads to the following basic procedure in dealing with claims under Sub-Clause 4.12 (Unforeseeable Physical Conditions).

- In the preparation of his claim submittal, the Contractor must first demonstrate unforeseeability (by an experienced contractor), with particular reference to Sub-Clause 2.5 (Site Data and Items of Reference) and Sub-Clause 4.10 (Site Data), and any other data that may be contained elsewhere in the Contract Documents.
- The unforeseeable condition must be encountered at the Site. Unforeseeable conditions off-site do not meet this criterion.
- The unforeseeable condition must be physical and not concerned with administrative events for example.
- Adverse *'climatic conditions on the Site, such as effect of rainfall'*, wind or abnormal temperatures are excluded.
- Adverse hydrological conditions, such as flows of water, are admissible including those attributable to off-site conditions such as flooding from a nearby stream or river.

A significant portion of claims submitted under this sub-clause relate to sub-surface geological conditions, which may require expert opinion in support of the claim. Claims relating to adverse hydrological conditions require an evaluation of the statistical frequency and severity of the unforeseeable event and a demonstration that the frequency and/or severity of the event was not foreseeable by an experienced contractor. The cost of this type of unforeseeable event may be covered in part by contract insurances, but it should be borne in mind that insurers do not award extensions of time.

Certain extreme categories of natural disasters which could also not been foreseen by an experienced contractor are included in Sub-Clause 19.1 (Definition of Force Majeure).

Should the Contractor consider that he has encountered an Unforeseen Physical Condition, he is required to give Notice to the Engineer in accordance with the 28-day period stated in Sub-Clause 20.2. The Insurers should also be notified of the event without delay.

The Contractor has a duty to continue with the Works regardless of his claim that he has encountered an Unforeseen Physical Condition.

The early FIDIC Guide states that the Contractor *'is expected to use his expertise'* to over-come the adverse conditions. The Engineer should cooperate with the Contractor to iden-tify technical solutions which fulfil the principles of the Engineer's design. This cooperation is important because remedial work or changes to the performance of the Works may in themselves represent Variations and Adjustments as described in Clause 13.

Should the Engineer for whatever reason decline to participate in the process to find solutions, the Contractor may have to proceed unilaterally. In such case, it is vital that he keeps the Engineer informed of the Contractor's proposals which should be supported by adequate technical documentation.

In dealing with the perceived Unforeseen Physical Condition, the Contractor must main-tain detailed records of his activities on a day-by-day basis and present them to the Engineer for agreement.

The Contractor is entitled to evaluate any future claim based on those records. DAAB and Arbitration boards may take a negative view of any non-cooperation.

Significant costing problems can arise if major items of Contractor's equipment and other resources are randomly used for work relating to the Unforeseen Physical Condition and to routine contract work. Separate cost control headings for each activity should be maintained.

In the event of a valid claim for Unforeseen Physical Conditions, Sub-Clause 4.12, the Contractor is entitled to payment of (additional) Costs incurred in overcoming the Unforeseen Physical Condition. The Contractor is entitled to payment for the original work at the billed rates (or varied rates) and, in addition, payment of the cost of any additional measures necessary to deal with the Unforeseen Physical Condition. Payments at bill rates include a profit allowance, whereas payment of cost excludes profit.

In practice it may be difficult, if not impractical, for the Contractor to divide the total package of work affected by the Unforeseen Physical Condition into a component of work directly related to the Unforeseen Physical Condition and separately into a component of original work.

The Contractor may elect to value the total package on the basis of Cost, but would then lose the profit element on the original work component. In preparing his records, the Contractor should consider the possibility of separating the two components. Again, this is a topic that could be usefully discussed in advance with the Engineer.

The First Edition introduced a proviso concerning more favourable conditions which is retained in this Second Edition. Before any additional cost is finally agreed or determined, the Engineer may (permissively) review whether other physical conditions in similar parts of the Works (if any) were more favourable than could reasonably be foreseen when the Contractor submitted the Tender. It may be presumed that the criterion of foreseeability applies to that which could be foreseen by an experienced contractor and does not refer to what the Engineer considers foreseeable.

The above is most likely to apply to projects involving repetitive work – building founda-tions, machine bases, and similar. Should the Engineer determine that the Contractor has received a benefit because of more favourable conditions, then the cost due elsewhere to the Contractor for proven unforeseeable physical conditions shall be reduced accordingly. This process shall not result in a net reduction in the Contract Price.

Finally, the Engineer, in making his determinations, may take account of the physical conditions foreseen by the Contractor when submitting the Tender. The earlier FIDIC Guide noted that if a dispute arises and is referred to the DAB or to arbitration, the members may wish to view evidence of the Contractor's assumptions, query the authors, and query why this evidence was not provided to the Engineer at an earlier date.

4.13   Rights of Way and Facilities

This sub-clause is essentially unchanged from that stated in the First Edition 1999.

Sub-Clause 2.1 of these Conditions of Contract requires the Employer to give the Contractor *'right of access to, and possession of, all parts of the Site. . .'*.

However, it may happen that the Contractor requires other rights of access to the Site in addition to those provided by the Employer under the terms of the Contract. The Contractor is required obtain such additional rights of way at his own risk and cost.

In addition, the Contractor may wish to occupy areas not within the Site and not otherwise within areas to be provided by the Employer under the terms of the Contract. Typically, these could be areas for quarries, borrow area, and additional storage areas. Again, the Contractor is to obtain these areas at his own risk and cost. Restrictions on the usage of public roads by construction traffic also require investigation.

4.14   Avoidance of Interference

This sub-clause is essentially unchanged from that stated in the First Edition

'The Contractor shall not interfere unnecessarily or improperly with (a) the convenience of the public and (b) the access to and use of all roads and footpaths. . .'.

The Contractor shall indemnify the Employer against all damages, losses, and costs which may arise under this heading. Third Party Insurance cover is to be provided by the Contractor in accordance with Sub-Clause 17.4.

4.15   Access Route

This sub-clause is largely unchanged from that stated in the First Edition excepting that new wording has been added, which states that if the non-suitability or non-availability of an access route arises because of design changes authorised by the Employer, then the Contractor would be entitled to payment of any additional costs arising and to an extension of time.

Most public authorities have strict rules concerning the use by contractors of roads falling under their authority and the needs of the Contractor will have to be negotiated in some detail.

Difficulties may arise on contracts for the construction or reconstruction of public roads. The Engineer, on behalf of the Employer, will routinely provide details of any road junctions and any other public access routes. However, difficulties frequently arise in respect of private dwellings located along the road whose access routes are severed by the road construction. The Engineer is required to provide detailed instructions of access to these roadside buildings (if any). The Contractor is entitled to payment for any work instructed.

During their pre-tender visit, the Contractor's tender personnel should review the issue of access to the site in general terms supplemented by preliminary discussions with the appropriate authorities.

4.16 Transport of Goods

Note:
The heading of this sub-clause does not include reference to the inter-related subject of customs arrangements. Guidance on customs arrangements is to be sought elsewhere in the Contract Documents.

'Plant' is defined in Sub-Paragraph 1.1.65 as equipment, machinery etc. intended to form part of the Permanent Works. It is assumed that any Plant to be provided by the Contractor will be identified in the Specifications and/or the Bill of Quantities. The Employer may himself provide items of Plant under separate contracts with other contractors or suppliers. In this eventuality the Contract Documents should identify any involvement of the Contractor and how any services provided will be paid for.

'Goods' are defined in Sub-Paragraph 1.1.44 as Contractor's Equipment, Materials, Plant and Temporary Works. These will not be incorporated in the Permanent Works. It is important for the Contractor to know if any Plant provided by him, and any Goods are subject to imposition of local taxes and/or customs duties. Considering the administrative difficulties which would arise if this were allowed, it is likely any concession allowed will be limited to supplies of fuel and cement which may be under direct government control.

Most Goods are likely to be sourced from outside the country of execution of the Contract. The Employer must therefore clearly indicate if the Contract is to have a 'tax free' status and/or if temporary imports are subject to payment of 'deposit payments' to be reimbursed on re-export.

On completion of the Contract, it is likely that the Contractor will want to re-export the majority of his plant and equipment, excluding any items the Employer may wish to purchase and scrap-only items. Therefore, ideally, the Contractor would prefer a temporary tax-free importation permit for those items which will be re-exported.

Materials and other minor items are likely to be consumed or otherwise scrapped and permanent tax-free importation is required.

Concerning item (b) of this sub-clause: to facilitate inspection, an organised contractor will arrange for permanent materials to be separated in his warehousing arrangements.

Concerning item (c): customs clearance is most likely to be arranged by a specialist clearing agency with support from the Contractor.

Concerning item (d): most items ordered by the Contractor will be shipped free on board (f.o.b.). A separate shipping insurance may be provided but the insurances required for the transport to site onwards are likely to be part of the Contractor's general All-Risk insurance policy.

4.17 Contractor's Equipment

Contractor's Equipment brought on the Site is deemed to be exclusively intended for the execution of the Works and shall not be removed from the Site without the Engineer's consent.

The Contractor is required to give a Notice of the date on which any major item of equipment arrived on Site. Subcontractor equipment is to be separately identified. Frequently, subcontractor equipment is required for haulage of materials such as cement, crushed rock, gravel, etc. from off-site locations, as not all Contractor's equipment will be suitable for use on public roads.

Occasionally, the off-site use of specialised Contractor's equipment may be requested by the Employer or other parties. Should the Contractor be willing or able to assist the requesting party, he should request the Engineer to provide his agreement for the temporary removal from Site. The need for appropriate insurances has to be addressed.

4.18 Protection of the Environment

The corresponding sub-clause of the First Edition has been slightly expanded in this Second Edition.

The Contractor is required to:

- protect the environment on or off the Site
- comply with any environmental impact statement which may be assumed to be included in the Contract Documents
- limit damage and nuisance to people and property resulting from pollution, noise etc.

An additional paragraph has been added, requiring the Contractor to ensure that emissions, surface discharges, etc. shall not exceed the values indicated in the Specification or those described by applicable laws. It is suggested that a suitable member of the Contractor's laboratory staff be trained and given charge of this subject.

4.19 Temporary Utilities

This sub-clause was titled 'Electricity, Water and Gas' in the First Edition

This sub-clause provides for two possibilities:

(a) The Contractor will be responsible for the supply of all utilities (notably including telecommunications) which the Contractor may require for the execution of the Works. There is no stated obligation for the Contractor to provide facilities for the use of other parties excepting that which is indicated in Sub-Clause 4.6 (Co-operation).

  The Contractor will need to determine whether he provides his own facilities by his own means, e.g. generators, water treatment plant, or whether it is possible to connect to local service providers.

(b) The Employer may himself arrange for the availability of the required facilities to a convenient distribution point on the Site, from which the Contractor may extend to suit his work requirements. The Contractor would be required to pay for the use of these services.

  In the long term, this has advantages for the Employer, since following the handover of the Works by the Contractor he is left with functioning service arrangements.

4.20 Progress Reports

It has long been a requirement of FIDIC contracts that the Contractor prepares monthly progress reports for submittal to the Engineer.

The First Edition provided for the first time a detailed listing of eight main topics to be addressed in these monthly reports. The same eight topics with small modifications are included in the Second Edition

There is one notable amendment. The First Edition Sub-Clause 4.21 item (a) included a provision for data provided by each nominated Subcontractor to be included in the monthly progress prepared by the Contractor.

This wording has been deleted in the Second Edition. However, this Second Edition, Sub-Clause 5.1 (Subcontractors) states that *'the Contractor shall be responsible for the work, for managing and coordinating all Subcontractors' work. . .'*. The presumption is that the Contractor is to continue to include Subcontractor data in the Monthly Progress Report.

The listing of topics to be included in the Monthly Report is lengthy and multifaced and is most suited to a major international project, including a significant content of Plant (mechanical and electrical) which is likely to be produced outside the country of where the contract is to be performed. It is not suitable for a relatively small routine road project where the language of the contract is not the language of the country, adding to difficulties in preparing the report. However, it is noted that Sub-Clause 4.21 starts with the wording *'Unless otherwise stated in the Particular Conditions . . .'*, opening the possibility that the listing of topics to be included in the Monthly Report can be modified to best suit the needs of individual projects.

Consequently, it is important that the Particular Conditions are reviewed to ascertain what modifications, if any, have been made to the listing of topics provided in the General Conditions.

The detailed requirements are onerous and will require careful planning by the Contractor in order that these reports shall be prepared as scheduled.

The format and content of any report does not have to be agreed or approved by the Engineer. However, it would be appropriate if the Engineer and Contractor were to informally review the general format of the report to ensure that it was fit for purpose.

The first report is to cover the period of the first calendar month following the Commencement Date. Thereafter reports are to be submitted monthly and shall continue to be submitted until the Date of Completion of the Works (or until any outstanding work is complete). Each report is to be provided within seven days after the last day of period covered by the report. Allowing for non-work days, this means that the Contractor has approximately 5 days to complete and submit each report. However, this is the same period of the monthly cycle that the Contractor is required to finalize and submit his Application for Interim Payment (Sub-Clause 14.3).

It is likely that the Contractor's Representative will delegate prime responsibility for preparation of the Monthly Reports to a suitably qualified member of his staff. This individual is to ensure that other members of the Contractor's staff provide suitably collated data for inclusion in the report. Much of the required data will already be compiled by various departments on the Contractor's organisation for internal use.

It is noted that Sub-Clause 14.3 (Application for Interim Payment) requires that the application is accompanied by a further copy of the Progress Report.

This Sub-Clause 4.20 (Progress Reports) concludes with a statement stating '. . . *nothing stated in a progress report shall constitute a Notice under a Sub-Clause of these Conditions*'. Should the progress report contain reference to an event which may give rise to a claim, the Contractor should give a separate Notice referenced to Sub-Clause 20.2 (Claims for Payment and/or EOT).

4.21   Security of the Site

This sub-clause was numbered as Sub-Clause 4.22 in the First Edition and has the same wording.

There are many elements to be provided and controlled by an effective security system. These requirements will vary significantly according to the location of the Works (urban, countryside, isolated etc.); the Contract may be executed at a number of locations etc. The list of topics to be evaluated is seemingly endless.

Health and Safety requirement of the Contract also must be taken into account. Security staff have to be identified and trained. Many of these items will need to be in place at the commencement of operations. Consequently, the services of a company specialised in security matters may be required.

4.22   Contractor's Operations on Site

This sub-clause was numbered as Sub-Clause 4.23 in the First Edition. The wording of the sub-clause has been slightly changed but the obligations of the Contractor are unaltered.

The first paragraph remains unaltered from that given in the First Edition. The Contractor is to '*confine his operations to the Site, and to any additional areas which may be obtained by the Contractor. . .*'.

The second paragraph requires the Contractor to keep the Site free from all unnecessary obstructions and to remove from Site all waste and items no longer required. Waste materials taken to a registered dump may incur charges.

The third paragraph states that *'promptly'* after receiving a Taking-Over Certificate, the Contractor shall remove his equipment, waste materials etc. from Site, leaving the Works in a clean and safe condition.

At the date of the issue of a Taking-Over Certificate much of the Contractor's Equipment is likely to have been removed from Site to another more convenient location, particularly so if it is to be re-exported.

4.23   Archaeological and Geological Findings

This sub-clause was numbered as Sub-Clause 4.24 in the First Edition under the heading "Fossils". Items of interest are to be secured by the Contractor and a Notice describing the finding is to be sent to the Engineer who shall issue instructions for dealing with it.

Items of geological or archaeological interest found on the Site shall be handed to the Employer (for the Engineer to hand on to the Employer). The Contractor shall also give a Notice to the Engineer, describing the finding and requesting instructions for dealing with it.

If the Contractor suffers delay or incurs costs complying with the Engineer's instructions, then, subject to Sub-Clause 20.2, the Contractor shall be entitled to EOT and payment of Costs.

The FIDIC Guide comments that *'this sub-clause makes no reference to the finding (of fossils) having to be unforeseeable because the Contract should specify the procedure in respect of foreseeable findings'.*

Although this sub-clause requires that any findings are to be placed in the care of the Employer, the Contractor should be aware of any requirements of the local law.

5

Clause 5 Subcontracting

Clause 5 of the First Edition was headed 'Nominated Sub-Contractors' which implied none of the sub-clauses which followed related to subcontractors who are directly engaged by the Contractor.

5.1 Subcontractors

In contrast, this opening sub-clause of the Second Edition refers to both Nominated Subcontractors and the Contractor's own subcontractors hereafter referred to as a 'in-house Sub-Contractor'. Sub-Paragraph 1.1.78 defines a 'Subcontractor' as *any person appointed by the Contractor as a subcontractor for a part of the Works. . .'*.

The Contractor shall obtain the Engineer's prior consent to all proposed subcontractors, except for the following:

 (i) Suppliers of materials. Effectively, the Contractor is obliged to obtain the Engineer's consent to all in-house subcontractors
(ii) All nominated subcontractors named in the Contract.

Sub-Clause 5.1 also states that the Contractor is *'responsible for the Work of all Subcontractors'* and *'for the acts or defaults of any subcontractor. . . as if they were acts or defaults of the Contractor'*.

The above requirements represent a significant additional risk placed on the Contractor. By definition, a nominated subcontractor is likely to be a specialist engaged in the manufacture of complex machinery or equal which is outside the Contractor's technical knowledge. As written, the *'acts and defaults'* of a Nominated Subcontractor could include design faults, delayed shipment of components, faulty assembly and testing by the nominated subcontractor etc. The Contractor is expected to manage any interface work between him and the nominated subcontractor and provide assistance to him on request.

In comparison, the Contractor is always responsible for work of his in-house subcontractors.

There are limitations on the value of works which the Contractor may sub-let to in-house subcontractors (refer to Contract Data for details) and there may be parts of the

Guide to the FIDIC Conditions of Contract for Construction: The Red Book 2017, First Edition.
Michael D. Robinson.
© 2023 John Wiley & Sons Ltd. Published 2023 by John Wiley & Sons Ltd.

Works which the Contractor will not allowed to sub-let and which he is obliged to execute himself.

Where the Contractor intends to engage in-house subcontractors, he shall submit the necessary details to the Engineer for his consent to engage any individual subcontractor. The Engineer is required to approve or disapprove the Contractor's request within a period of 14 days of receiving the request, failing which, he is deemed to have approved the Contractor's request.

Thereafter the Contractor is required to give Notice of the intended commencement date of the in-house subcontractor and the date of his commencement of work on Site.

5.2 Nominated Subcontractors

5.2.1 Definition of a 'Nominated Subcontractor'

In addition to the definition given in Sub-Paragraph 1.1.78, it is stated in this Second Edition that *'nominated subcontractor means a Subcontractor named as such in the Specification or whom the Engineer under Sub-Clause 13.4 (Provision Sums) instructs the Contractor to employ as a subcontractor'*. This definition corresponds to the definition provided in the First Edition.

5.2.2 Objection to Nomination

This sub-paragraph provides an opportunity for the Contractor to object to a nominated Subcontractor proposed by the Engineer. A listing of reasons which would entitle the Contractor to make a valid objection are provided in the text.

Most of these reasons can be dealt with by the appropriate wording of the subcontract document.

However, item (a) is crucial: *'. . . if there are reasons to believe that the Subcontractor does not have sufficient competence, resources or financial strength'*.

These matters may be discussed in a meeting between Engineer, Contractor, and proposed Subcontractor. Essentially, Contractors are wary of subcontractors who are not able to act independently and who rely on access to the resources of the Contractor for which payment will have to be made. Secondly, considering the contents of Sub-Clause 5.1, the Contractor will need to be assured that the proposed Subcontractor(s) are fully aware of their duties and obligations under the Main Contract.

5.2.3 Payments to Nominated Subcontractors

This Sub-Paragraph corresponds to Sub-Clause 5.2 of the First Edition.

Sub-Clause 14.3 (Application for Interim Payment) describes how the monthly Application for Interim Payment will be compiled and presented to the Engineer for certification. The value of the work performed (and payable) by the nominated Subcontractor(s) will form part of the Application for Interim Payment.

Once payment is received, the amount due to the Subcontractor(s) is to be paid as soon as possible. The exact detail is to be provided in the Subcontract agreement(s).

In addition to the above, the Subcontractor(s) may have taken goods and services from the Contractor for which payment is due. A normal commercial invoice should be provided by the Contractor to the Subcontractor for settlement. These matters should not become intermingled with payment under the (main) Contract.

5.2.4 Evidence of Payment

Any subcontract agreement between the Contractor and a nominated Subcontractor is to include details of the nominated subcontractor's bank to which the Contractor shall make payment of the amounts due (following receipt of these amounts from the Employer).

A mechanism whereby the Subcontractor formally acknowledges receipt of payment should be established (and copied to the Engineer if necessary).

Note:
Should the Contractor make claims against the nominated Subcontractor for events solely due to the failings of a nominated Subcontractor or vice versa, which are not due to events falling under the main Contract, these will have to be settled directly between the Contractor and the Subcontractor and paid for without the involvement of the Engineer.

Should the Subcontractor make a claim which fall under the (main) Contract, then the claim must be resolved by reference to the Engineer, following the procedures given in the (main) Contract.

6

Clause 6 Staff and Labour

6.1   Engagement of Staff and Labour

During the pre-tender period, the Contractor's tender preparation team will visit the general area where the project will be executed and assess the local labour market and related issues. Local knowledge will be obtained from the appropriate Government ministry and local labour offices. The Contractor may already have a local office or partner who will assess the project requirements.

Concerning accommodation, there are many possibilities:

- Local workers may be able to live in their own homes and be transported to work by the Contractor. This is always a better option for social reasons than living in purpose-built barracks close to the workplace.
- Expatriate staff and senior local staff may be largely accommodated in existing housing where available. If travel distances are too great, then a purpose-built village may be the better option.

6.2   Rates of Wages and Conditions of Labour

The Rate of Wages and Conditions of Labour are to be in conformity with local law. There is a strong probability that there will be a local trade union with whom the Contractor will need to negotiate wages and conditions.

6.3   Recruitment of Persons

The Contractor and the Employer may not attempt to recruit staff and labour from each other. However, as the project draws to a close, the Employer may have a long-term interest to recruit some of the Contractor's staff for operation and maintenance duties with which they will already be familiar.

Guide to the FIDIC Conditions of Contract for Construction: The Red Book 2017, First Edition.
Michael D. Robinson.
© 2023 John Wiley & Sons Ltd. Published 2023 by John Wiley & Sons Ltd.

6.4   Labour Laws

The Contractor is to comply with the relevant labour laws on all matters. Expatriate staff will require work permits enabling them to reside and work in the country. At the conclusion of the Contract, the expatriate staff are to be returned to their country of origin.

6.5   Working Hours

(a) This sub-clause opens with the wording 'No work shall be carried out on the Site on locally recognised days of rest or outside the normal working hours stated in the Contract Data unless stated in the Contract'.

Working practices will vary from project to project and from country to country. For example, a road rehabilitation project executed in Eastern Europe will likely adopt a conventional 5-day working week. In contrast, a hydro-electric project at an isolated site in Africa will likely adopt a 6-day working week (not least because the Employer would require the resultant electric power supply at the earliest date). The actual permitted working hours are to be found in the Contract Data (Page 4).

(b) On major projects certain operations are continuous – examples include tunnel excavation/lining, formwork/concreting in two shifts, crusher operations and transport of materials (subject to local restrictions). Security and safety staff are required to be continuously available together with medical care. The Engineer is required to give his consent to these types of operations.

(c) It may happen that work may have to be performed in an emergency situation. The Contractor is to give immediately Notice to the Engineer, describing the emergency and the remedial or recovery work required.

6.6   Facilities for Staff and Labour

The Contractor shall provide all necessary accommodation and welfare facilities for his personnel. The location of these facilities shall be agreed with the Employer. The Contractor is not required to provide facilities for the Employer's Personnel and others which, by definition, includes the Engineer and his staff.

6.7   Health and Safety for Personnel

'*The Contractor is required at all times to take reasonable steps to maintain the health and safety of his personnel*', this basic requirement almost certainly being far more explicitly stated in local laws and regulations.

The Contractor is required to ensure '*that medical staff, first aid facilities, sick bay and ambulance services are available at all times at site*'. The precise requirements will vary considerable from project to project, from location to location, from the size and complexity of the site operations and not least from the amount of support available from local hospitals and health services. Should a sizeable part of the Contractor's staff be expatriate in origin, evacuation plans to return injured or sick expatriates to their own country may also be required.

As a first step, it is necessary to meet with local medical professionals to ascertain what facilities are locally available and how they might be accessed by the Contractor. This will have an impact on decisions taken in respect of medical facilities and staff to be provided by the Contractor on Site.

It may be expected that the Engineer will require the Contractor to provide a detailed proposal to cover all medical aspects of health and safety. Sub-Clause 6.7 further requires the Contractor to make all necessary welfare arrangements and to be responsible for hygiene.

Welfare arrangements in this context would include messing and feeding arrangements, the provision of drying rooms and similar. Hygiene requirements would include adequate toilet and washing facilities and a general regard for the preventative measures against disease and unhealthy work practices.

The Contractor is required '*to appoint a health and safety accident prevention officer at the Site, responsible for maintaining health, safety and protection against accidents*'. The person shall be required to be qualified for this work and to be given authority to issue instructions and take protective measures.

These general requirements can be expected to be amplified in the Contract and will be subject to the relevant laws of the country.

Most major contractors will have available formal Health and Safety Manuals. The accident-prevention officer will need to supplement these manuals to conform with the requirements of the Contract and the local law. Parts of the safety manuals may require translation into the local language(s). Periodic reporting is required and will form part of the Progress Report described in Sub-Clause 4.20.

6.8 Contractor's Superintendence

In staffing an overseas project, the Contractor may be assumed to already have in his employ a number of key staff and workers with suitable experience gained on similar projects. Additional staff can be recruited 'by word of mouth' or more general advertising. The CVs of the key staff will be available with the Contractor's Representative.

6.9 Contractor's Personnel

The Contractor's Personnel in question will be hourly paid local staff. Disruptive personnel will be dealt with according to the Employment Laws of the country.

6.10   Contractor's Records

The occupations and actual work hours of personnel and the type and actual work hours of plant will be recorded on time sheets prepared by 'time clerks' stationed in each sector of the works.

6.11   Disorderly Conduct

Local disturbances on site can be dealt with by site security staff. More serious disturbances may require the intervention of the local authorities.

6.12   Key Personnel

The opening sentence of this sub-clause states *'If no Key Personnel are stated in the Specification, this sub-clause should not apply'*. Note also the reference to this Sub-Clause 6.12 in the Contract Data (Page 30). The organisation of the Contractor's Key Staff is best dealt with as part of Sub-Clause 4.3.

7

Clause 7 Plant, Materials, and Workmanship

7.1 Manner of Execution

This sub-clause (which is identical to that included in the First Edition) serves as a reminder to the Contractor of the expectations of the Employer. Should the Employer have significant doubts with regard to the ability of a potential contractor to execute the Work to the required level of expertise then these concerns should be expressed during any pre-award meetings.

7.2 Samples

This sub-clause has been significantly expanded from that given in the First Edition

The provision of samples by the Contractor and his suppliers for testing and confirmatory testing by the Contractor in cooperation with the Engineer fall into several categories:

(1) Aggregates for concrete or asphalt or for use as filler material
 – locally obtained from quarries or borrow
 – sampled and tested in Site laboratory
(2) As for (1) but produced by off-site commercial quarries
 – Initial testing by producer
 – Confirmatory sampling and testing on Site by Contractor
(3) Imported re-bar and steel sections
 – Manufacturer's test certificates to be provided
 – Confirmatory site tests and sampling as required by the Engineer
(4) Re-bar and steel sections obtained from local suppliers acting as agents for foreign suppliers
 – Test certificates by foreign supplier to be provided (fake certificates are not uncommon)
 – Sampling for on-site check testing required

Guide to the FIDIC Conditions of Contract for Construction: The Red Book 2017, First Edition. Michael D. Robinson.
© 2023 John Wiley & Sons Ltd. Published 2023 by John Wiley & Sons Ltd.

(5) Products of all sorts provided and certified by manufacturer as conforming to specified standards.
(6) Machinery and Plant Manufacturer may be a general manufacturer, 'in-house' subcontractor to the Main Contractor, or a nominated subcontractor or supplied (and fixed) by another contractor.

    Items under this heading would include:
- Transformers, Excitation cubicles, Gates, Valves, etc.
- Motors, Transport Cranes, Weighbridges, etc.

    These items will be manufactured off-site (and very likely in another country). Exceptional items may require the Engineer to attend the assembly plant. Sampling requirements to be clarified either in the Contract or otherwise arranged by the Engineer acting on behalf of the Employer.

7.3 Inspections (by the Employer's Personnel)

1. Engineer is to have access to all places on site from which natural materials are being obtained.
2. The Engineer should have full access to check the progress of manufacture of Plant and production and manufacture of Materials. The authority of the Engineer under this sub-head is clear, but, as noted in the commentary on Sub-Clause 7.2, this may be complex in certain circumstances.

Good housekeeping by the Contractor's storekeeper is important for safe and appropriate storage of Plant and Materials intended for inclusion in the Works (the so-called 'Permanent Material'). Where practical, a section of the store area is to be reserved for these items, enabling the Engineer to carry out any inspections he may require.

These items must not be removed for any purpose other than for the specific usage noted in the purchase order and store card.

The general inspection of the Site will be a continuous duty of the relevant Engineer's Staff who are entitled to use the accesses, walkways, pathways, ladders, scaffolding, etc. provided by the Contractor for his own (and subcontractors') staff.

The Contractor is required to give formal Notice to the Engineer that work items are ready for his inspection. Prime instances would be inspection prior to commencing a concrete operation or that a work area is ready for commencement of a filling operation. Present convention is that the relevant permission forms are prepared and approved by the Engineer's and Contractor's field staff directly supervising these operations.

The Engineer may have to make special arrangements for inspection/approvals to be carried out at a foreign location. The services of specialist testing agencies, such as TÜV, Lloyds, may be required.

These requirements (if any) should be clarified in these Conditions of Contract.

7.4   Testing by the Contractor (on Site)

The Contractor is required to provide a furnished, fully equipped laboratory, supervisory staff and assistants with adequate transport, etc. as required, so that he may perform all types of testing specified in the Contract.

Any requirements of the corresponding Engineer's staff are not given in these General Conditions, however, it would be mutually convenient that the Engineer has site offices close to the laboratory, as this would facilitate the process of the Contractor giving a Notice to the Engineer in respect of specified testing of materials.

The requirements of this sub-clause apply to all tests specified in the Contract excluding testing described in Clause 9 (Tests on Completion). The wording *'specified testing'* can cover a wide range of different types of testing, ranging from testing of major items of installed Plant which may require the services of external specialist engineers and technicians, to routine continuous testing of concrete, aggregates, soils, etc. which can be expected to be continuous testing performed in suitably equipped site laboratories or directly in the workplace by site-based staff.

A delay in the testing programme caused by the Contractor may cause the Engineer to incur standing costs in respect of his non-resident specialist staff. If the planned testing programme is disrupted by the absence of the Engineer's staff, then the Contractor will incur standing costs in respect of his non-resident specialist staff. In these circumstances, the Party responsible for the cause of delay would be responsible for the additional costs of the other Party.

In contrast, site testing of concrete, aggregates, etc. is a continuous process routinely performed by the resident staff of both Engineer and Contractor. Should one activity be disrupted, these staff members will continue with other activities. The monthly cost of these staff members is not affected by the amount of work performed. It is unlikely that either Party could demonstrate that additional staff costs were incurred.

7.5   Defects and Rejection

This sub-clause is an expansion of the corresponding sub-clause of the First Edition. The stated sequence of events is:

- The Engineer gives Notice to the Contractor of any defects or otherwise (It would be hoped that the relevant staff of the Engineer and Contractor would discuss and arrange the correction of the majority of defects without formality).
- The Contractor prepares a proposal for remedial work for the approval of the Engineer.
- Within 14 days of receipt of the Contractor's proposal, the Engineer may accept or reject, asking for a more detailed proposal. If not rejected, the Contractor can commence work immediately.
- If the defects persist, the work will be re-tested and the cycle repeated.

Should re-testing be required, necessitating the attendance of the Engineer, then '*... the Employer shall be entitled, subject to Sub-Clause 20.2, to payment of these costs by the Contractor*'.

7.6 Remedial Work

This sub-clause refers to remedial work to be undertaken by the Contractor before a Taking-Over Certificate is issued.

(i) The Contractor is required to repair or remedy or replace Plant and Materials not in accordance with the Contract.
(ii) The Contractor is to repair or remedy any other work not in accordance with the Contract.
(iii) The Contractor is to carry out remedial work which is urgently required for the safety of the Works.

The Contractor shall bear the cost of all remedial work excepting for remedial work which is necessitated by an act of the Employer or the Employer's Personnel or by an Exceptional Event (defined in Sub-Clause 18.1). The Contractor is entitled to EOT and Payment of Cost plus Profit in respect of both these events.

Should the Contractor fail to comply with the Engineer's instructions in respect of any remedial work, the Employer may employ another party to carry out the repair work. Except to the extent the Contractor would have been entitled to payment for the remedial work, the Employer, subject to Sub-Clause 20.2, would be entitled to payment by the Contractor for all costs arising from the failure.

7.7 Ownership of Plant and Materials

Plant (defined in Sub-Paragraph 1.1.65) and Materials (defined in Sub-Paragraph 1.1.53), both of which are intended to form part of the Permanent Works, become the property of the Employer when delivered to Site.

The value of the Plant and Materials delivered to Site shall be recovered by the Contractor in the next IPC (refer to Sub-Clause 14.5). This value is reduced according to the value of the Plant and Materials consumed in each period.

It may be important to establish the legal ownership of Plant and Materials, particularly in the event of bankruptcy/liquidation of the person who is in possession of them. For example, some plant may be owned by Subcontractors, or be the subject of a hire agreement or a hire-purchase agreement. This is a complex topic. Local law will prevail, and legal advice would be necessary to protect the rights of the Contractor and the Subcontractors and creditors.

Reference may be also made to Sub-Clauses 14.3(a) and 14.5, which provide for interim payments to be made to the Contractor for Plant and Materials intended for the Works and delivered to site – all subject to specified conditions.

7.8   Royalties

The Contractor is responsible for payment of royalties and rents for natural materials obtained from outside the Site. Rents are to be paid to the owner of the quarry/borrow and royalties are taxes to be paid to the government.

In addition, the Contractor is responsible for any costs arising from the use of disposal sites outside the Site. It may be possible to dispose of uncontaminated surplus excavated material by careful landscaping.

Domestic waste and industrial waste also are to be disposed of in accordance with the local laws. Fees may become due.

These matters should be discussed with the Employer during the official site visit prior to tender date.

8

Clause 8 Commencement, Delays, and Suspension

8.1 Commencement of Works

The Engineer is required to give a Notice to the Contractor stating the Commencement Date not less than 14 days before Commencement Date.

The Engineer will only be authorised to issue such a Notice when instructed by the Employer to do so, implying that the Employer has issued a Letter of Intent to the Contractor, requiring him to commence mobilisation. The signing of the full Contract Agreement will follow later.

In some jurisdictions, the concept of a Letter of Intent is not legally acceptable, so there may be delays whilst copies of the full Contract Agreement are prepared for signature. These requirements may have a delaying effect on the date on which the Engineer issues a Notice to Commence.

With the contractual formalities being finalised, including the Engineer's Notice to Commence, the Contractor should insist on a formal inspection of the Site, to ensure that there are no additional obstructions to commencement, including continuous occupation of the Site by third parties, often as a consequence of delayed expropriation. The access road to the Site may have been inspected at the time of the official inspection but now must be fully available to the Contractor.

As soon as the precise Commencement Date is known, the Contractor can now activate his insurances accordingly (reference: Clause 19 (Insurances)).

8.2 Time for Completion

The wording of the corresponding sub-clause of the First Edition has been slightly modified in the Second Edition without changing the intent of the sub-clause.

'The Contractor shall complete the whole of the Works, and each Section within the Time for Completion for the Works or Section'.

The cross-reference to Sub-Clause 10.1 (Taking Over the Works and Sections) is maintained and it is to that sub-clause that the detailed requirements for the Employer's Taking Over are provided.

Guide to the FIDIC Conditions of Contract for Construction: The Red Book 2017, First Edition.
Michael D. Robinson.
© 2023 John Wiley & Sons Ltd. Published 2023 by John Wiley & Sons Ltd.

8.3 Programme

This sub-clause has been significantly expanded and re-written in the Second Edition

The opening sentence reads *'The Contractor shall submit an* <u>initial</u> *[underscoring added] programme for the Works to the Engineer within 28 days after receiving the Notice under Sub-Clause 8.1 (Commencement of the Works)'.*

The programming software to be used shall be stated in the Specifications. Otherwise, the programming software shall be acceptable to the Engineer. Depending on the scale and complexity of the Works to be executed, the Employer may have provided in the Tender documents for the guidance of tenderers an outline programme of the Works to be executed.

Consequently, the Contractor, in preparing his tender, may elect to use the programming software used by the Employer in preparing the outline programme.

8.4 Advance Warning

This is a new sub-clause, not included in the First Edition.

Each Party is required to inform the other Party of any existing or anticipated future events which may affect the performance of the Works. It is quite probable that the Parties will already be aware of these events; however, it would be prudent for the Contractor to give a formal Notice to the Engineer with specific reference to this sub-clause.

The following categories of events are identified and are topics which could be discussed in Site Meetings (Sub-Clause 3.8 (Meetings)):

(a) <u>events which adversely affect the work of the Contractor's Personnel</u>
Considering that the definition of Contractor's Personnel (Sub-Paragraph 1.1.17) includes staff, labour, and other employees, the word 'event' could be widely interpreted to range from civil disturbances, strikes, to epidemics etc.

(b) <u>events which adversely affect the performance of the Works when completed</u>
Examples could include the adverse effect of faulty design, Plant not performing as intended, etc.

(c) <u>events which increase the Contract Price</u>
Since the word 'event' is not defined, the scope of this item is not clear. The Contract Price will increase by the valuation of variations, settlement of Contractor's claims, the value of works performed in excess of those quantities provided in the Bill of Quantities. Therefore, it is prudent that the likely final value of the Works is re-evaluated at appropriate intervals, so that current and projected increase in the final value of the Works is available to both Parties and the Engineer.

(d) <u>events which delay the execution of Works or Section</u>
Delays may either be the responsibility of the Contractor, in which case he is obliged to rectify the situation, or delays which fall under the responsibility of the Employer (variations, additional work, Exceptional Events, etc.).

The Engineer may request the Contractor to submit proposals for mitigation of these events.

8.5   Extension of Time

This sub-clause corresponds to Sub-Clause 8.4 of the First Edition but has been modified in many respects.

Causes entitling the Contractor to an Extension of Time are listed as follows:

(a) <u>A Variation</u>

This item also states that *'there shall be no requirement to comply with Sub-Clause 20.2 (Claims for Payment and EOT)'*. However, Sub-Clause 20.2/20.2.1 requires the Contractor to give a Notice to the Engineer if he considers himself entitled to additional payment. A problem arises because the Contractor may not be informed for some time of the manner in which a claim will be dealt with by the Engineer – a variation or a claim settlement. Consequently, it is recommended that the Contractor ignores the additional wording quoted above and gives a formal Notice to the Engineer on every occasion he considers himself entitled to additional payment or Extension of Time.

(b) <u>A cause of delay giving an entitlement to EOT under a sub-clause of these Conditions</u>

(c) <u>Exceptional adverse climatic conditions</u>

Two items require reviewing by the Contractor to support a claim made under this heading:

(i) Historical climatic records obtained from relevant official weather stations if not provided in the Contract Data.

(ii) Climatic data obtained from a weather station on the Site (possibly located adjacent to the Site Laboratory), where climatic records may be recorded and maintained for future reference.

(d) <u>Unforeseeable shortages in the availability of Personnel or Goods</u>

Shortages of skilled labour is a frequent problem. This may be alleviated to some extent by training the work force 'on the job'. Potential skill shortages should be identified at the pre-tender stage, so that a training regime can be quickly arranged. A check should be made to ascertain if skilled workers from other countries can be brought to the Site, even on a short-term basis.

The Contractor must be able to demonstrate that he has made every effort to overcome potential labour shortages.

Local supply of basic materials including cement, fuels, reinforcing steel, can quickly be disrupted by events not within the Contractor's control, even should the Contractor have reasonable stocks of these materials.

Detailed records of the above are to be maintained and the Engineer kept informed. As stoppages become unavoidable, a Notice should be sent to the Engineer, detailing the delays and additional cost accruing.

(e) <u>Delays, impediment or prevention by the Employer or Employer's Personnel</u>

In addition to the items included in the above listing, the Contractor is entitled, subject to Sub-Clause 20.2, to EOT, if the measured quantity of any work is greater than the estimated quantity of this item in the Bill of Quantities by more than 10% and such increase in quantities causes a delay to completion for the purposes of Sub-Clause 10.1 (Taking Over the Works and Sections).

However, the agreement or determination of any such claim (Sub-Paragraph 20.2.5) may include a review by the Engineer of measured quantities of other items of work which are significantly less (by more than 10%) and determine any favourable effect on the critical path of the Programme.

When determining each EOT, '. . . the Engineer shall review previous determinations under Sub-Clause 3.7 and may increase but not decrease the total EOT'.

In respect of concurrent delays caused by Employer/Contractor, the text refers to the *'rules and procedures stated in the Special Procedures'*. The guidance is given in Sub-Clause 8.5 of the Special Procedures (Page 34). However, the guidance acknowledges that *'. . . there are no one standard set of rules/procedures in use internationally'* and *'different rules/procedures may apply in different legal jurisdictions'*. Compromise may be required in order to reach a fair solution.

8.6 Delays Caused by Authorities

Provided the Contractor has diligently followed the procedures of the public authorities, utility companies and these authorities delay or disrupt the Contractor's work and the delay or disruption was unforeseeable, then this delay or disruption will be considered a cause of delay under Sub-Clause 8.5 (Extension of Time).

Although there is no mention of a Notice to the Engineer (Sub-Clause 20.2), it would be appropriate to issue such a Notice for the avoidance of doubt. It will be noted that this sub-clause does not itself provide for payment of the Contractor's costs.

The assistance of public authorities and private utility is frequently required by the Contractor to provide various services, including:

- the provision of electricity, water, communication facilities to the Site and other locations (new services)
- relocation or repair of an existing service
- land expropriation issues
- general access and use of public roads

It may happen that the authority may not have the necessary finance, equipment, or manpower to provide the requested services. Consequently, the Contractor may be requested to assist the public authority by supplying materials and use of Contractor's equipment in order that the execution of the Works is not delayed.

Although not stated in this sub-paragraph, should the progress of the Works be disrupted, it is appropriate to give a Notice to the Engineer when these obstructions arise, with a claim of Extension of Time and/or payment of additional Costs incurred (referencing Sub-Clause 20.2).

8.7 Rate of Progress

Progress and Production will always be a key topic for discussion in the Contractor's internal management meetings. Key staff, including senior site staff, workshop manager, programmer, and a technical office manager should be among the attendees. The causes of any obstructions causing delay or deviations are to be identified.

- Was Site mobilisation too slow? If so, can lost time be recovered by providing additional resources?
- Is the execution of primary activities below the production levels on which the programme of work is based? (e.g. earth moving quantities, quarry production, concrete plant capacity, asphalt plant production etc.)?
- Is one activity under-resourced and under-producing and therefore disrupting the following activities? If so, additional resources available for deployment?
- Do the field supervisors have a clear understanding of both short-term and long-term targets?

Should remedial measures be put in place, they must be seen to be effective. Actual progress must be continuously monitored.

Sub-Clause 8.7 also states that if actual progress is too slow to complete the Works or Section within the Time for Completion and/or progress is not in accordance with the Programme, then the Engineer may instruct the Contractor to submit a revised programme describing the revised methods which the Contractor proposes to introduce to ensure completion in accordance with the Contract. Due allowance shall be made for any EOT entitlement due to the Contractor. Unless the Engineer issues a Notice to the Contractor to the contrary, the Contractor shall proceed in accordance with the revised programme and methods at his own risk and cost.

Sub-Clause 8.7 also states *'the revised methods to expedite progress may include increases in working hours'*. The provisions of Sub-Clause 6.5 (Working Hours) states in part *'No work shall be carried out on the Site. . . .. outside the normal working hours stated in the Contract Data. . .'*. Large projects may require multi-shift continuous 24-hour operations, whereas the standard working hours on an urban contract may be strictly restricted by local laws to a single shift/five days a week operation. These issues require clarification in the Contract Data.

If the revised method of working causes the Employer to incur additional costs (presumed to indicate additional supervision costs), then the Employer shall be entitled to recover those additional costs from the Contractor subject to Sub-Clause 20.2.

This sub-clause concludes by stating that revised methods, including acceleration methods instructed by the Engineer to reduce delays listed under Sub-Clause 8.5 (Extension of Time for Completion) shall be evaluated in accordance with Sub-Paragraph 13.3.1 (Variation by Instruction).

In some circumstances, it may be difficult to determine if the acceleration is a consequence of delays, which are the responsibility of the Contractor, or a consequence of delays, which are the responsibility of the Employer.

8.8 Delay Damages

This sub-clause is a repeat of Sub-Clause 8.7 of the First Edition, excepting that an additional sentence has been added, which states that the Contractor's liability for Delay

Damages shall not be limited in case of fraud, gross negligence, deliberate default or reckless misconduct. Otherwise Delay Damages are limited to the amounts stated in the Contract Data. The Employer cannot claim his actual costs but does not have to demonstrate his actual costs or losses.

The Contractor cannot prevent the imposition of Delay Damages by submitting late or fictious claims for Extension of Time. However, the Employer may lose his entitlement to claim delay damages if he prevents a valid Extension of Time being agreed or determined in accordance with Sub-Clause 20.2.

Should the Employer intend to claim Delay Damages, he is required to present a documented claim as provided for in Sub-Clause 20.2. The Engineer is required to formally consult with the Parties before making a determination.

8.9 Employer's Suspension

The heading of this sub-clause in the First Edition was 'Suspension of the Works'. The amended title makes it clear that the Contractor does not have a formal role to play in the ordering of a suspension of the Works, although doubtlessly he will make representations to the Engineer if he considers a suspension is required or inevitable.

It is the Engineer who instructs the Contractor to suspend the Works. Considering the significance of any suspension of the Works, it may be presumed that the Engineer has discussed the need for a suspension with the Employer in advance of any instruction issued to the Contractor. The Contractor has a general obligation to protect the Works (or parts thereof) against loss or damage (refer to Clause 9 (Insurance)).

The following Sub-Clauses 8.10, 8.11, and 8.12 are applicable only if the event giving rise to the issuance of a suspension order is within the contractual responsibility of the Employer.

However, it may be that the responsibility for the event leading to the suspension is not immediately clear. In such a situation, the Contractor should proceed as described in Sub-Clauses 8.10, 8.11, and 8.12 until responsibility is decided. The co-operation of the Engineer to agree records should be sought accordingly.

8.10 Consequences of Employer's Suspension

If the Contractor suffers delay and/or incurs Cost from complying with the Employer's Suspension, he is entitled to an Extension of Time and/or payment for any Cost incurred plus profit.

Two exclusions are noted:

- The consequences of Contractor's faulty or defective workmanship, Plant or Materials (i.e. Permanent Materials). Note: this exclusion refers only to items which are '. . .*faulty or defective*. . .'. Should the Employer's suspension cause damage or loss to items which are (or were) not defective, the Employer would be responsible for the consequences (time and money).

- The failure of the Contractor to protect against loss or damage. It may not be feasible to provide full protection in certain circumstances. The Contractor must make every effort to provide protection but cannot be responsible for losses in extreme circumstances.

In complying with an instruction to suspend and protect the Works, the Contractor will inevitably incur costs. The Site, generally, and particularly static equipment, will have to be secured; mobile equipment will need to be brought to a secure area, and most of the work force will be placed on stand-by or released, etc. A detailed status of the Site will have to be recorded and agreed with the Engineer on (at least) a daily basis. The evaluation of the Contractor's Costs will not be a simple task and if the Contractor does not have an effective cost office on Site, it will be difficult and take time to achieve. The Contractor would also be entitled to including a portion of both Site and home office overheads in his total costing. The Contractor would also be entitled to claim for lost profit.

The entitlement of the Contractor to EOT can only be quantified once the suspension is lifted, but Costs and EOT entitlement will continue in part until all of the Contractor's operations are working at the rate existing before the suspension was instructed.

The Contractor is not entitled to EOT or payment of Cost incurred

- if the suspension arises from a cause which is the responsibility of the Contractor
- if the suspension is the consequence of defective workmanship, Plant and Materials provided by the Contractor.

Further, the Contractor has no entitlement in respect of the Contractor's failure to fully protect the Site and all assets located and stored on Site. The level of protection to be provided should be discussed and agreed with the Engineer at the earliest opportunity.

8.11 Payment for Plant and Materials after Employer's Suspension

The Contractor is entitled to payment of the <u>value</u> at the date of suspension for Plant and Materials which have not been delivered to Site if

(a) the work on Plant or delivery of Plant and Materials has been suspended for more than 28 days and
   (i) the Plant and Materials were programmed to have been completed and ready for delivery to Site
   (ii) the Contractor provides the Engineer with reasonable evidence that the Plant and Materials comply with the Contract
(b) the Plant and Materials are marked as the Employer's property.

In practical terms the word 'value' could be interpreted as the amount of the relevant supplier invoices for Plant and Materials ordered by the Contractor to which must be added transport costs, insurance costs and customs/clearing charges.
   or
If the Plant and Materials are provided by a Nominated Subcontractor, the value may be calculated from original invoices (in a similar manner adopted by the Contractor noted above) or by reference to a priced listing in the sub-contract documents.

It is not clear if value is subject to deductions and adjustment (refer to Sub-Clause 14.3), including the Contractor's mark-up on subcontract work.

8.12 Prolonged Suspension

If the suspension exceeds 84 days, the Contractor may give a Notice to the Engineer requesting permission to proceed. If the Engineer does not respond within 28 days of the Notice, the Contractor may either:

- agree to a further extension entitling the Contractor to Extension of Time and payment of his continuing Costs plus profit
or
- after giving a second Notice, treat the suspension as an omission of the affected part of the Works, immediately releasing the Contractor from any obligation to protect, store or secure the Employer's property
or
- if the suspension affects the whole of the Works, the Contractor may give a Notice of Termination under Sub-Clause 16.12.

In an extreme situation, a period of 112 days could elapse before the Contractor is advised whether the Works are to continue or if the period of suspension is to be extended by agreement or if the Works are to be terminated. The Contractor is placed at considerable commercial risk by any suspension and the uncertainty arising from an extended suspension amplifies those risks.

It is important for the Contractor to maintain close contact with the Employer and the Engineer in order to obtain an agreement concerning the future of the Contractor's resources both on and off Site. Good record-keeping by the Contractor is essential to support any claims subsequently presented by him.

8.13 Resumption of Work

The Contractor shall resume Work as soon as practical after receiving a Notice from the Engineer to re-commence the suspended work.

The Contractor and the Engineer have an obligation to jointly examine the affected Works and the Engineer shall record their findings.

A lengthy suspension of Work may affect more work areas than the original affected parts of the Site and these should be noted in the Engineer's report. It is possible that elements of the Contractor's work force may have been laid off and are to be recalled. The affected parts of the Site require re-activation and made ready to commence normal operations. These restorative activities require recording as they constitute additional Contractor Cost, requiring reimbursement.

The Works and the Plants and Materials shall be inspected for deterioration, loss, or damage, which is to be made good by the Contractor. According to Sub-Clause 8.10, the Contractor is responsible for any Costs arising as a consequence of protective measures employed. The Costs of these protective measures are not foreseen in the Contractor's tender, are not a Contractor's risk item and must be included and recovered as part of the Costs to be reimbursed under Sub-Clause 8.9, Employer's Suspension, or Sub-Clause 8.10, Consequences of Employer's Suspension.

The so-called protective measures to be provided should ideally be agreed between the Contractor and the Engineer.

9

Clause 9 Test on Completion

9.1 Contractor's Obligations

This sub-clause opens with a reference to Sub-Clause 7.4 (Testing by the Contractor), which serves as a reminder of all those preparatory matters to be addressed before any testing can begin, notably the provision of calibrated equipment and availability of experienced personnel to perform the testing.

The bulk of the testing to be performed is likely to be centred on the testing of electric and mechanical Plant and Equipment and Communication Equipment, all with their connective systems.

The Contractor is required to prepare a comprehensive testing programme for presentation to the Engineer at least 42 days before the intended start date of the test programme. Nominated Subcontractor input, if any, will also be required. The Engineer may review the proposed test programme and give a Notice to the Contractor, identifying items of non-compliance. Within a further 14 days, the Contractor shall present a revised test programme to the Engineer. The Contractor shall not commence the test programme until a Notice of No-objection is given by the Engineer.

In practice it is customary for the concerned staff of the Contractor and Engineer to hold a series of working meetings with the objective of agreeing the detail of a test programme. Effectively, there will be a number of test programmes, each covering a different part of the total testing programme. The attendance of the Employer should be welcomed, particularly if he is to be the operator/maintenance agency after the Taking-Over is satisfactorily completed. It is also possible that the Employer will have an input on the testing programme for on-load testing.

The sub-clause then specifies that the Contractor shall give a Notice to the Engineer of the date after which the Contractor will be ready to carry out each of the Tests on Completion. The Contractor shall commence the Tests on Completion within 14 days after this date.

The results of the Test Programme(s) will be recorded as the testing programme proceeds and signed by the Contractor or Subcontractor staff. A copy is to be provided to the authorised Engineer's staff member present during the performance of the testing. A formal document will be provided to the Engineer for each group of type of test.

Guide to the FIDIC Conditions of Contract for Construction: The Red Book 2017, First Edition.
Michael D. Robinson.
© 2023 John Wiley & Sons Ltd. Published 2023 by John Wiley & Sons Ltd.

9.2   Delayed Tests

Having given due Notice to the Engineer of an intention to commence testing, should the testing be unduly delayed by the Engineer's personnel (or another party for whom the Engineer is responsible), then the issue will be dealt with in accordance with Sub-Clause 10.3. Sub-Clause 10.3 entitles the Contractor to recovery of any additional Costs and EOT in accordance with Sub-Clause 20.2.

If Tests on Completion are delayed by the Contractor, the Engineer may give Notice to the Contractor, requiring him to carry out the testing within a further period of 21 days after receipt of the Notice. Should the Contractor continue to delay the Tests on Completion, then the Engineer may give the Contractor a second Notice. If the second Notice is not acted upon by the Contractor, the Engineer may carry out the tests himself. Copies of the test results are to be sent to the Contractor by the Engineer. The Employer is entitled to recovery of additional costs incurred because of delayed testing for payment by the Contractor, all subject to Sub-Clause 20.2.

Note:
These Conditions of Contract do not refer to the possibility of the Engineer operating and testing affecting a supplier's warranty.

9.3   Re-testing

If the Works (or a Section) fail to pass the Tests on Completion, the Engineer or the Contractor may require these failed tests to be repeated on the same terms and conditions.

9.4   Failure to Pass Tests on Completion

This sub-clause refers to the possibility that the Works (or a Section) may fail to pass the Tests on Completion following re-testing as described in Sub-Clause 9.3. It may be assumed that the Contractor and his suppliers/subcontractors and nominated subcontractors will be aware of the circumstances of the repeated failure and will make every effort to correct the causes of failure.

The following options are then open to the Engineer:

(a) The Engineer may order a further repetition of Tests on Completion
(b) The Engineer may reject the Works if the effect of the failure is to deprive the Employer of substantially the whole benefit of the Works. The options available to the Engineer are stated in Sub-Clause 11.4 (Failure to Remedy Defects).
(c) Should the continuing failure relate to a Section of the Works, the options available to the Employer are also provided in Sub-Clause 11.4.
(d) The Engineer may issue a Taking-Over Certificate if requested by the Employer. The Contractor shall proceed in accordance with the execution of the Works. The Employer is then entitled to claim his costs from the Contractor (Sub-Clause 20.2 refers) or a reduction in the Contract Price (Sub-Clause 11.4b refers).

10

Clause 10 Employer's Taking Over

10.1 Taking Over of the Works and Sections

The Works shall be taken over by the Employer when the Works have been completed in accordance with the Contract including the Tests on Completion (refer to Clause 9 (Tests on Completion)).

The basic procedure for the Employer's Taking Over given in the First Edition has been retained: *'The Works shall have been completed in accordance with the Contract, including the passing of the Tests on Completion.*

The Engineer shall issue the Taking-Over Certificate excepting for any minor outstanding works, and defects which will not substantially affect the use of the Works . . . for their intended purpose'.

In this Second Edition, three additional conditions have been added:

1. *'if applicable, the Engineer has given . . . a Notice of No-objection to the as-built records submitted under Sub-Paragraph 4.4.2 (As-Built Records)'.*
2. *'if applicable, the Engineer has given . . . a Notice of No-objection to the operation and main-tenance manuals under Sub-Paragraph 4.4.3 (Operation and Maintenance Manuals)'.*
3. *'if applicable, the Contractor has carried out the training under Sub-Clause 4.5 (Training)'.*

The First Edition did not specify that items 2 and 3 above were conditions to be fulfilled before the Engineer's Taking-Over Certificate could be issued. Consequently, in the past, Contractors have tended not to give these activities a high precedence.

The Contractor may apply for a Taking-Over Certificate by giving a Notice to the Engineer not more than 14 days before the Works will be complete and ready for taking over.

Consequently, as the anticipated date for Taking Over draws nearer, it is recommended that the Contractor maintains an increasingly detailed dialogue with the Engineer and continuously review the status of the Works, thereby establishing the extent of the Works yet to be completed.

A 'punch list' of outstanding work items should be prepared for review at appropriate intervals with completed items marked as complete. Such a continuous dialogue would greatly facilitate the Taking Over process.

Guide to the FIDIC Conditions of Contract for Construction: The Red Book 2017, First Edition.
Michael D. Robinson.
© 2023 John Wiley & Sons Ltd. Published 2023 by John Wiley & Sons Ltd.

However, in some jurisdictions, it may be unlawful for the Engineer to issue a Taking-Over Certificate without obtaining the consent of the Employer. Matters can be yet more complicated if the Employer in turn is required to hand over to an end user who may not be specifically mentioned in the Contract Document. Such bureaucratic processes can severely delay the actual date of the Works for reasons not of the Contractor's making.

The Contractor, preferably in cooperation with the Engineer, who may also not be familiar with local procedures, should seek authoritative guidance at an early stage of the execution of the Contract.

Sub-Clause 10.1 continues by stating that if the Engineer fails to issue the Taking-Over Certificate or fails to reject the Contractor's application within the stated period of 28 days, and if the Works are indeed substantially completed in accordance with the Contract, the Taking-Over Certificate shall be deemed to have been issued on the last day of that period. It seems inevitable that the Employer will insist on the Engineer rejecting the Contractor's application without due cause or with improper reasoning rather than allow the Taking Over to take place by default. Additionally, the provisions of local law as described earlier may make such Taking Over procedure illegal.

The Contractor should be very wary that he does not find himself trapped in a situation where a delay can be engineered between the Employer, the end user, and the authorities, preventing the rightful issue of the Taking-Over Certificate, in order to meet the convenience of the Employer rather than complying with the stated intentions of the Contract. The Engineer may find his authority undermined in these circumstances. It further leaves the Contractor in a difficult position because it is not possible to estimate the time to be allowed in the programme for completion of this activity.

The Employer has no right to use the Works if the Contractor has failed to complete them in accordance with the Contract except after Termination.

The issue of the Taking-Over Certificate for the Works or Sections results in a reduction of the amount of Delay Damages. The value of the Delay Damages shall be reduced proportionately by the value that the works taken over bears to the Accepted Contract Amount.

10.2 Taking Over of Parts of the Works

'The Engineer may, at the sole discretion of the Employer issue a Taking-Over Certificate for any part of the Permanent Works'.

Should the Employer require access to (and possibly operate) parts of the Works prior to the issue of a Taking-Over Certificate, then such requirement either must be stated in the Contract or regularised in a separate agreement made between the Parties.

Whilst the Employer and Contractor may make a supplementary agreement permitting the Employer to occupy parts of the Works without formal Taking Over, in practice this should be used sparingly. Not only are there plenty of opportunities for disputed liability for any damage caused to already completed work and problems with insurers, but also there is the fact that the Employer is obtaining a principal benefit in advance which may unreasonably delay the eventual issue of the Taking-Over Certificate.

However, in road projects, especially rehabilitation projects, it may be unavoidable to prevent use of completed sections of the road before formal Taking Over, particularly if lengthy diversions are not permitted or not feasible.

Should the Employer take possession before a Taking-Over Certificate is issued:

- the part occupied by the Employer is deemed to have been taken over from the date the Employer takes possession. Again, this may not be in conformity with the local law. In addition, there may be political considerations to be evaluated.
- The Contractor is no longer responsible for the care of the part of the Works taken into the possession of the Employer. The standard Contractor's All Risk Insurance Policy will almost certainly exclude liability for any damage or loss caused by the Employer's actions. The Contractor should be wary of falling between the requirements of the Contract and the exclusions of the insurers.
- *'If requested by the Contractor, the Engineer shall issue a Taking-Over Certificate for this part of the Works'*. No procedure is provided, and it may be assumed that an appropriate letter of request would be sufficient to require the Engineer to act. Again, this procedure may not be in accordance with local law. It is not in the interest of the Contractor to have the Employer in possession of a part of the site when the assumed handover procedure does not conform to the procedures required by the local law.

The issue of a Taking-Over Certificate for a part of the Works results in a reduction of the amount of Delay Damages. The value of the Delay Damages shall be reduced proportionately by the value that the works taken over bears to the Accepted Contact Price.

Should the Contractor incur cost as a result of the Employer taking over and using a part, the Contractor shall be entitled, subject to Sub-Clause 20.2, to payment of such Cost plus Profit. Notice of Claim required.

10.3 Interference with Tests on Completion

'If the Contractor is prevented for more than 14 days (either a continuous period, or multiple periods which total more than 14 days) from carrying out the Tests on Completion by the Employer's Personnel or by a cause for which the Employer is responsible, then:

(a) *the Contractor shall give a Notice to the Engineer describing such prevention*
(b) *the Employer shall be deemed to have taken over the Works (or Section) on the date when the Tests on Completion would otherwise have been completed*
(c) *the Engineer shall immediately issue a Taking-Over Certificate as soon as practical.*

Finally, the Engineer shall give Notice to the Contractor of not less than 14 days after which the Contractor can recommence the outstanding tests'.

During the execution of the Tests on Completion the following typical situation could arise which prevents the Contractor from completing the test programme:

The Contractor has satisfactorily completed all tests on completion including static tests on hydraulic equipment installed by him, but the final load test requires the provision of the final connections to power and water supply which is a responsibility of the Employer, as he is relying on another contractor or public utility company to provide the final connections. It is inevitable that the final load testing will be delayed, and the Contractor will be unable to complete the Tests on Completion on schedule.

If the Contractor suffers delay and incurs Cost as a result of being prevented from carrying out the Tests on Completion, the Contractor shall be entitled, subject to Sub-Clause 20.2, to EOT and/or payment of such Cost plus Profit.

10.4   Surfaces Requiring Reinstatement

As part of site clearance and removal of surplus materials, the Contractor will be required to reinstate surfaces to their original condition or to a condition acceptable to the Employer.

It is possible that the surplus materials may be disposed of in a manner beneficial to the Employer.

Typically, any work to be carried out under this heading, if not complete at the date of Taking Over, will be included in a list of outstanding works to be attached to the Taking-Over Certificate.

11

Clause 11 Defects after Taking Over

This sub-clause has been re-titled from 'Defects Liability' used in the First Edition. The formal identification of defects is the subject of Sub-Clause 7.5 (Defects and Rejection) and the general procedures for the correction of any defects is given in Sub-Clause 7.6 (remedial Work). This Clause 11 overlaps with Sub-Clauses 7.5 and 7.6 and is centred on other aspects of the remedying of defects.

11.1 Completion of Outstanding Work and Remedying Defects

The term 'Defects Notification Period' (DNP) is defined in Sub-Paragraph 1.1.27 as *'the period for notifying defects in the Works. . .'.* The period is given in the Contract Data and is calculated from the date of issuance of the Taking-Over Certificate (Sub-Clause 10.1 refers).
During the Defects Notification Period the Contractor is required to:

- complete any work outstanding on the Date of Completion. However, the wording *'within a reasonable time as instructed by the Engineer'* used in the First Edition 1999 has been enlarged to *'within the Times stated in the Taking-Over Certificate or such other reasonable time as instructed by the Engineer'.* This implies that the Contractor will have an input into the timing of the execution of any repairs. This requirement reinforces the need for the Contractor (in co-operation with the Engineer) to prepare a 'punch list' of outstanding works, including any remedial works, ahead of the likely Date of Completion in order that the list of outstanding work is kept to a minimum
- execute all remaining outstanding work and repair before the expiry date of the Defects Notification Period. Assuming a Defects Notification Period of twelve months, it is recommended that the Engineer and Contractor inspect the Works at approximately three-monthly intervals and identify those outstanding works completed and if new defects have arisen, within each time interval. At this stage of the Contract, the Contractor will have significantly reduced his site establishment in accordance with the requirements of Sub-Clause 11.11 (Clearance of Site). Consequently, outstanding works (if any) and repair works should be completed expeditiously to ensure that the Date for Completion is not extended.

Guide to the FIDIC Conditions of Contract for Construction: The Red Book 2017, First Edition.
Michael D. Robinson.
© 2023 John Wiley & Sons Ltd. Published 2023 by John Wiley & Sons Ltd.

The Contractor is not responsible for damage to the Works caused by the Employer or others permitted to use the Works by the Employer, including the general public. However, the Contractor remains responsible for any damage caused by his own employees and subcontractors completing any activities during the Defects Liability Period. It is important in this period that the Contractor works in an orderly and controlled manner in order that his employees do not cause further damage while carrying out their tasks.

The satisfactory completion of any outstanding work and the rectification of defects and repairs should be formally recorded and signed-off by the Parties. Periodic joint inspections of the Works by the Parties during the Defects Liability Period would be beneficial, particularly for larger and more complex projects.

11.2   Cost of Remedying Defects

The Contractor is responsible for all costs arising out of his obligation to correct defects and repair damages that are within his contractual responsibility. However, should the Contractor consider that the repair works are attributable to other causes, not the responsibility of the Contractor, then he shall give a Notice to the Engineer accordingly. The Engineer is required to *'agree and determine'* the cause of the defect. The date of the Notice by the Contractor is the date of commencement of the time for limit for agreement under Sub-Paragraph 3.7.3 (Time Limits).

In this Second Edition, the Contractor is responsible for other defects or damages arising from:

- design (if any) of the Works for which the Contractor is responsible
- Plant, Materials, or workmanship not being in accordance with the Contract
- improper operation or maintenance attributable to errors or inadequacies in As Built Records or operational manuals, or damages arising from inadequate training of the Employer's staff.

There are other issues which may cause damage to the Works:

- Damages arising on the Site as a consequence of breaches of site security are likely to be the Contractor's responsibility.
- Damages arising from the actions of other contractors engaged by the Employer are the responsibility of the Employer (although the matter may directly be settled by the insurers of the other contractors).
- Damages caused to the parts of the Site by the Public. Occasionally the Employer may request the Contractor to allow public usage of part of the Site prior to the issuing of a Taking-Over Certificate. This situation may particularly arise on a lengthy road project and should be discussed and agreed between the Parties before implementation. One solution is that a Taking-Over Certificate for part of the Works could be issued.
- However, the Contractor remains responsible for any damage caused by his own employees and subcontractors completing any activities during the Defects Notification Period. It is important in this period that the Contractor works in an orderly and controlled manner in order that his employees do not cause further damage while carrying out their tasks.

In preparing a Notice to the Engineer the Contractor should carefully review the contractual basis of his claim.

The satisfactory completion of any outstanding work and rectification of defects and repairs should be formally recorded and signed-off by the Parties. Periodic joint inspections of the Works by the Parties during the Defects Notification Period would be beneficial, particularly for larger and more complex projects.

11.3   Extension of Defects Notification Period

It may happen during the Defects Notification Period that the Works or Section or Part cannot be used as intended due to a defect or damage. The Employer is entitled to an extension of the Defects Notification Period (DNP) corresponding to the period of non-serviceability. The extension is limited to the defective item. The Defects Notification Period is not to be extended for a period of more than two years after the DNP stated in the Contract Data.

11.4   Failure to Remedy Defects

If the Contractor fails to remedy any defect or damage notified in Sub-Clause 11.1 within a reasonable time, a date may be fixed by the Employer by which the defect or damage is to be remedied. A Notice of this fixed date shall be given to the Contractor by the Employer.

If the Contractor fails to remedy the defect or damage by the notified date and, provided this remedial work was to be executed at the cost of the Contractor, then the Employer may:

(a) carry out the work himself or by other contractors at the Contractor's cost. The Employer is entitled to reimbursement of his costs in accordance with Sub-Clause 20.1 (Claims)
(b) accept the damaged or defective work, in which case the Employer shall be entitled to a reduction in the Contract Price in accordance with Sub-Clause 20.1 (Claims)
(c) require the Engineer to treat any part of the Works which cannot be used for its intended purpose as an omission as if it were instructed under Sub-Clause 20.1 (Claims)
(d) terminate the Contract with immediate effect. This is self-evidently a very extreme situation and will rarely occur. Legal advice would be essential if such a situation were to occur.

11.5   Remedying of Defective Work off Site

If the defect or damage cannot be remedied on Site, the Contractor may remove the defective items for repair off site with the consent of the Employer. The Contractor is to give a Notice explaining the reasons why the defective items require removal off site and shall provide full details of the intended removal for the approval of the Employer.

In giving consent to the removal, the Employer may require the Contractor to increase the amount of the Performance Security by the full replacement cost of the defective or damaged Plant.

If the defective or damaged Plant was provided by Subcontractors (either the Contractor's in-home subcontractor or a Nominated Subcontractor), then the Contractor may be entitled to reimbursement of the cost of increasing the value of the Performance Security by the Subcontractor.

Difficulties may arise if the Plant taken for repair off Site delays or interferes with the Employer's operations on a Section of the Works already taken over by the Employer. Temporary replacements may be required. It is important that the period in which the damaged plant is likely to be off Site is established before removal.

11.6 Further Tests after Remedying Defects

Within seven days of remedying any defect or damage, the Contractor shall give a Notice to the Engineer of any repeat testing which the Contractor intends to perform. Within a further seven-day period the Engineer is to give a Notice agreeing with the Contractor's proposal or instructing the repeat testing required.

Should the Contractor fail to respond to the proposals of the Engineer within 14 days of the remedial works, then the Contractor shall perform the repeat testing stated by the Engineer.

These tests shall be carried out at the risk and cost of the Party liable.

11.7 Right of Access after Taking-Over

Until the 28th day after the issue of the Performance Certificate the Contractor has a right of access to the Site for the purpose of rectifying defects and executing repairs. For security reasons, the Employer may impose controls and restraints on the Contractor's activities.

Whenever the Contractor requires access during the Defects Notification Period, the Contractor is required to give a Notice to the Employer giving the reasons for requiring access and the preferred date for access. Within seven days, the Employer shall respond accepting the Contractor's proposal or proposing an alternative date.

If there are unreasonable delays in obtaining access, the Contractor may, subject to Sub-Clause 20.2, claim for any additional Costs incurred plus Profit.

This sub-clause then continues to state that if the Employer fails to respond within seven days, the Employer is deemed to have given consent to the Contractor's proposals. This appears to be an unworkable solution should access be physically impossible and a discussion between the Employer and Contractor would provide a mutually better solution.

11.8 Contractor to Search

The Contractor shall, *'if required by the Engineer, search for the cause of any defect'* or by extension the circumstances of any damage. The Engineer is to direct this operation, which is to be performed on a date specified by the Engineer.

Should the Contractor be liable for the defect or damage, then the search will be carried out at his cost. If the Contractor is not responsible, then he is entitled to payment of his costs of search plus profit.

Should the Contractor fail to carry out the instructed search, then the Employer, by Notice to the Contractor, is entitled to specify a date on which the search will take place. The Contractor may attend at his own cost. If the defect is to be remedied at the Contractor's expense, the Employer is entitled to recover his costs from the Contractor.

11.9 Performance Certificate

The obligations of the Contractor under the Contract are not complete *'until the Engineer has issued a Performance Certificate'* (which shall state the date on which the Contractor completed his obligations).

The Performance Certificate shall be issued within 28 days after the latest of the expiry dates of the Defects Notification Period, provided the Contractor has supplied all Contractor's Documents and completed all testing of the Works.

'Only the Performance Certificate shall be deemed to constitute acceptance of the Works'. There is no formal action required by the Contractor other than to advise the insurers.

11.10 Unfulfilled Obligations

Notwithstanding the issue of the Performance Certificate there will remain unfulfilled obligations between the Parties. The most significant of these will relate to the settlement of financial matters and resolution of any remaining disputed issues.

The Employer is to return the Performance Security to the Contractor within 21 days of the date of the Engineer issuing the Performance Certificate.

11.11 Clearance of Site

Most contractors will remove their equipment, surplus materials, temporary works etc. from site as soon as possible. Alternatively, the Contractor may be allowed to use the Site (by agreement with the Employer) as a storage point pending removal of the Contractor's property to another project.

However, upon receiving the Performance Certificate, the Contractor is required to remove his remaining property within a further period of 28 days, unless agreed otherwise with the Employer. Failure to do so entitles the Employer to dispose of and sell the Contractor's property and to restore the Site. The Employer would be entitled to recover his costs and pay only the residual amounts to the Contractor.

Frequently, the Contractor may locate some part of his storage area outside the Site on land not belonging to the Employer. The Employer has no jurisdiction in respect of these external areas.

12

Clause 12 Measurement and Valuation

12.1   Works To Be Measured

This sub-clause in part states *'Whenever the Engineer requires a part of the Works to be measured on Site, he shall give a Notice to the Contractor . . . of the part to be measured and the date and place on Site where the measurement shall be made'*.

   and

   'Any part of the Permanent Works that is to be measured from records shall be identified in the Specification . . . such records shall be prepared by the Engineer'.

The Engineer is to give Notice to the Contractor's Representative when he requires any part of the Works to be measured and the Contractor is required to support and attend the Engineer. The Works are to be measured based on these records by the Engineer and agreed by the Contractor.

If the Contractor disagrees with the Engineer's records and/or measurement, he shall give Notice of his disagreement to the Engineer. The Engineer shall review the records and confirm or vary them. If the Contractor remains in dispute with the Engineer's records and/or measurement, he is obliged to give a second Notice of dispute to the Engineer within a further period of 14 days, otherwise the Engineer's records and/or measurement are considered accurate and correct.

There are two separate measurement processes to be considered:

(1) Interim measurements are required for the purpose of preparing the Application(s) for Interim Payment (Sub-Clause 14.3 refers).

   These interim measurements are routinely prepared on a monthly basis by land surveyors, quantity surveyors and others from both the Engineer's and Contractor's staff. These measurements can be modified in any future IPC but shall be as accurate as possible (with due recognition of the limited time available for the preparation of each monthly IPC). They do not constitute a final measurement since they are likely to be only a portion of the total of each bill item.

   However, in many instances, including work to be disturbed or covered up, or items of work which are complete but no longer accessible, can be measured as final measurements.

Guide to the FIDIC Conditions of Contract for Construction: The Red Book 2017, First Edition.
Michael D. Robinson.
© 2023 John Wiley & Sons Ltd. Published 2023 by John Wiley & Sons Ltd.

In practice the structured approach described in this Sub-Clause 12.1 is rarely adhered to. It is the Contractor who has the vested interest in ensuring that the measurement of work performed is maximised and, equally importantly, that is measured as quickly as possible after execution for inclusion in the next Interim Payment Application.

Whilst the Engineer may equally have the intention to measure the Works accurately and fairly, he is not under the same economic pressures as the Contractor. As a consequence, it is the Contractor who drives this measurement process and is prepared and able to commit resources to accomplish measurement as soon as practicable.

Significant portions of the measurement may be made from Drawings, but other operations such as major earthworks, particularly if construction is spread over a long period of time, may require several field surveys to be made as a basis for measurement.

Not all items of work are visible or accessible after their execution. In such cases, agreed measurement records must be progressively prepared and maintained. Equally, owing to site vagaries, additional operations may have had to be performed that cannot be later discerned from the As-Built Drawings.

Experienced Site staff will recognise the crucial need to maintain accurate records of measurement, which should be updated as appropriate.

Wherever possible and whenever items of work are completed, the measurement should be finalised and not left for agreement at some future, undefined date. Memories can be selective and unreliable. In addition, there may be changes in the relevant staff of both Engineer and Contractor, which may further delay finalisation of measurement. The Contractor should be particularly wary, since replacement Engineer's staff may not be receptive to agreeing earlier measurement not formally agreed by their predecessors. It is the Contractor who suffers financially and who is obliged to protect his interests by giving the issue of measurement its due priority.

It would be mutually beneficial for the Engineer and the Contractor to agree a standard format for recording agreed measurements prior to the commencement of Works.

In addition to the measurement of the physical works, there are usually a number of administrative bill items contained in the Contract, which are paid according to certified documentation to be provided by the Contractor. Paper items such as receipts, invoices, and records of numbers of vehicles provided for the Engineer and similar items have to be securely stored for future reference, particularly if the Contract is to be audited at a later date by third parties.

(2) Final measurement is a summation of the various parts or sub-parts of the Works executed in accordance with the Contract specifications. Some parts of this final measurement will have, effectively, been agreed in the preparation of IPCs as noted in 1) above and the remaining items progressively agreed as the Works progress to completion.

The actual amounts of work performed by the Contractor may not be the subject of a dispute between Engineer and Contractor. However, the Engineer may consider that the Contractor is not entitled to payment for all of the quantities of work actually performed, leading to a formal dispute.

If the difference between performed quantities and proposed paid quantities cannot be resolved, then the Contractor would be entitled to give a Notice to the Engineer and proceed with a claim in accordance with Sub-Clause 20.1 (Claims).

12.2 Method of Measurement

This Sub-Clause 12.2 has been revised from that given in the First Edition.
'The method of measurement shall be as stated in the Bill of Quantities'
and
'Except as otherwise stated in the Bill of Quantities, measurement shall be made of the net actual quantity of each item of the Permanent Works and no allowance shall be made for bulking, shrinkage and waste'.

It is an unfortunate fact that there does not exist an internationally recognised method of measurement. The method of measurement most frequently encountered is the Civil Engineering Standard Method of Measurement sponsored by the Institute of Civil Engineers London, but regrettably even this standard is not yet in widespread use on international contracts.

The Conditions of Contract are understandably somewhat vague on the issue of method of measurement and state only that *'measurement shall be made of the net actual quantity of each item'* and *'the method of measurement shall be in accordance with the Bill of Quantities or other applicable schedules'.*

It is unreasonable to assume that a contractor with international experience is necessarily familiar with local practices in any specific country. Unless he has staff member with appropriate local knowledge, he is, in the preparation of the Tender, going to be dependent on intelligent assessment of the Tender documentation. Given the limited time available for the preparation of the Tender, it will be difficult for a tenderer to make more than a brief check of the correctness of the quantities shown in the Bill of Quantities.

However, a tenderer should undertake a review of the descriptions included in the items in the Bill of Quantities, to check they are comprehensive and conform to the technical specifications. Any omissions or anomalies should be immediately formally queried with the Engineer.

Most unreasonably, it frequently happens that a Bill of Quantities will include a general preliminary statement to the effect that the individual bill item descriptions are not to be taken as a full description of the work included in the item. Sometimes this is interpreted by the Engineer to mean that for any work not specifically identified, but necessary because of work descriptions contained in the Specifications, the cost is deemed to be included in the individual bill item or spread over a group of bill items (refer to Sub-Clause 4.11). Further complications can arise when the Specifications contain misplaced references to the Bill of Quantities or in some cases contradict the descriptions in the Bill of Quantities. A tenderer has little option but to carefully review the wording of the Bill of Quantities and report any anomalies. In the absence of a standardised Method of Measurement, this remains an unsatisfactory process which hopefully will receive attention from FIDIC and the various trade organisations in the near future.

For the present the FIDIC Contracts Guide contains the following advisory considerations:

'(a) *If the Bill of Quantities includes principles of measurement which clearly require that an item be measured and if the Bill of Quantities does not contain such an item, then an additional Bill Item will be required.*

(b) *If the Bill of Quantities includes (either by reference or specified) principles of measurement which do not clearly require that a particular item of work be measured and the work was described in the Contract and does not arise from a Variation, then measurement does not require an additional Bill Item.*

(c) *If the Bill of Quantities does not contain principles of measurement for a particular item of work and the work was described in the Contract and did not arise from a variation, then measurement does not require an additional Bill Item'.*

Items (b) and (c) above emphasizes the need for the Contractor's estimating office to make a study to ascertain how and where each item of work will be paid for.

12.3   Valuation of the Works

The contents of this sub-clause have been re-written and expanded significantly from that included in the equivalent clause of the First Edition.

This sub-clause opens with the basic statement that the Engineer shall *'value each item of work by applying the measurement agreed or determined in accordance with Sub-Clause 12.1 and Sub-Clause 12.2 and the appropriate rate or price for the item. The appropriate rate or price for the item shall be the rate or price specified in the Bill of Quantities or if there is no such item, specified for similar work. Any item of work . . . . for which no rate or price is specified, shall be deemed to be included in other rates and prices in the Bill of Quantities'.*

The last sentence above is somewhat ambiguous. Does *'specified'* mean that the item is not to be separately valued, or does it mean that the item has in error not been priced by the Contractor in the preparation of his Tender?

To clarify the matter, the Contractor could query the intention in any Q and A exchange with the Engineer; or price the item fairly and have the issue clarified in any pre-contract review of the Tender offer.

A new rate or price is appropriate for any item of work if:

(a) no specific rate or price is appropriate because the item of work is not of similar character or is not executed under similar conditions as any item in the Contract

(b) (i)   the measured quantity of the item is changed by more than 10% from the quantity in the Bill of Quantities

(ii)   this change in quantity multiplied by the rate or price given in the Bill of Quantities exceeds by more than 0.01% of the Contract Amount

(iii)   this change in quantity directly changes the cost per unit quantity by more than 1%

(iv)   this item is not specified in the Bill of Quantities as a 'fixed rate item', fixed charge or similar term referencing to a rate or price that is subject to adjustment for any change in quantity

(c) the work is instructed under Clause 13 (Variations and Adjustments) and sub-paragraph (a) and (b) above applies.

Each rate shall be derived from any relevant rate or prices specified in the Bill of Quantities, considering the matters referred to in (a), (b) or (c) above. Otherwise, the new rate or price shall be based on the Cost of executing the Work with a profit margin of 5% added.

Should the Engineer and the Contractor not be able to agree the appropriate rate or price, then the Contractor shall give a Notice of the disagreement to the Engineer and shall proceed in accordance with Sub-Clause 3.7 (Agreement or Determination) to agree or determine the appropriate rate or price.

Until such time as the rate or price is agreed, the Engineer shall assess a provisional rate or price for the purpose of Interim Payment Certificates.

12.4 Omissions

12.4.1 Contractor to give Notice of unreimbursed Costs

'Whenever the omission of any work forms part (or all) of a Variation

(a) *the value of which has not been otherwise agreed*
(b) *the Contractor will incur (or has incurred) cost which, if the work had not been omitted would have been deemed to be covered by a sum forming part of the Accepted Contract Amount;*
(c) *the omission of the work will result in this sum not forming part of the Contract Price; and*
(d) *this cost is not deemed to be included in the valuation of any substituted work;'*

then the Contractor is required (sub-section (c)) of Sub-Paragraph 13.3.1 (Variation by Instruction), to give details to the Engineer accordingly with detailed particulars.

This sub-clause enables the Contractor to give a Notice to the Engineer of a Cost which has been incurred and which will not be reimbursed due to the omission of Work. The claim for reimbursement of these Costs shall be referenced to Sub-Clause 20.2 (Claims for Payment and/or EOT). The Engineer is required to make a determination in accordance with Sub-Clause 3.7 (Agreement or Determination).

13

Clause 13 Variations and Adjustments

The title of this clause 'Variations and Adjustments' is used in the First Edition and is retained in this Second Edition but significant changes have been made to the various sub-clauses.

A significant proportion of disputes have their origins in the formation and evaluation of Variations.

The Contractor is required to proceed expeditiously with the Works and yet is not authorised to vary the Works without first receiving appropriate instructions from the Engineer.

Sub-Clause 13.1 states that the Variations will be initiated by the Engineer and Sub-Paragraph 13.3.1 (Variations by Instruction) states that *the Engineer may instruct a Variation . . .*. The Employer is not formally involved in this process although it can be assumed that the Engineer will keep the Employer fully informed of actions taken. Should the Employer issue instructions directly to the Contractor, then the Contractor shall immediately refer the instruction directly to the Engineer for confirmation.

However, it may happen that the Particular Conditions of Contract contain wording which requires the Engineer to obtain the formal authorisation of the Employer before authorising a Variation (refer to the third paragraph of Sub-Clause 3.2 (Engineer's Duty and Authority)). Such a procedure is not in accordance with the General Conditions of Contract. This requirement may arise because project budget is limited or insufficient. Delays may occur whilst additional funding is obtained. Furthermore, Employers who are Public Authorities, may require the authorisation of higher authorities.

The resultant uncertainty can leave both Contractor and the Engineer having to deal with significant administrative uncertainties.

13.1 Right to Vary

Variations may be initiated by the Engineer under Sub-Clause 13.3 (Variation Procedure) at any time before the issue of the Taking-Over Certificate for the Works. However, it is important that the Variation be issued expeditiously and not allowed to extend the Time for Completion, ensuring that the issue of any Taking-Over Certificate is not delayed.

Guide to the FIDIC Conditions of Contract for Construction: The Red Book 2017, First Edition.
Michael D. Robinson.
© 2023 John Wiley & Sons Ltd. Published 2023 by John Wiley & Sons Ltd.

The Contractor is not obligated to carry out work not within the scope of the Contract or variation work instructed after issue of the Taking-Over Certificate. Often requests for such extra-contractual work arise close to the conclusion of the Works, as the Employer uses surplus funds to finance additional works not included in the original scope of the Contract. However, the Contractor may find it sufficiently profitable to carry out these works regardless of their contractual status.

Any varied work performed in the Defects Notification Period may require a revised pricing structure since the Contractor's cost base existing prior to the issue of the Taking-Over Certificate may no longer be applicable. This aspect requires a careful review by the Contractor without undue delay.

The Contractor shall be bound by and execute the Variation with due expedition and without delay. If the Contractor cannot comply, he shall give a Notice to the Engineer with detailed reasons, including:

(a) the Work was Unforeseeable. Having regard to the scope and nature of the Works described in the Specifications, the Contractor may not have the resources or the technical expertise to execute the varied works in good order.
(b) the required goods are not readily available
(c) The Contractor's liability in respect of Sub-Clause 4.8 (Health and Safety Obligations) and Sub-Clause 4.18 (Protection of the Environment) may be adversely affected.

These and any other issues are to be reviewed by the Engineer and instructions confirmed or varied.

'*Each variation may include:*'

(a) '*changes to the quantities of any item of work included in the Contract*'
    The Quantities of work included in the Bill of Quantities are estimates only. The final measured quantities of work may be different. Conventionally, changes in quantities that arise as a normal consequence of carrying out the originally planned works are measured within the existing bill items. Changes in quantities that arise as a consequence of a Variation are separately measured within the Variation.
(b) '*changes to quality and other characteristics of any item of work*'
    Typically, a change in the strength of concrete may be instructed which requires a change in cement content and aggregate grading. Other elements of costs including placement costs may be unaltered.
(c) '*changes to the levels, positions and/or dimensions of any part of the Works*'
    The quantities of work may be varied by these instructions. Some reworking is also a possibility.
(d) '*omission of any work*' (not to be performed by others)
    Reference is to be made to the valuation procedures described in Sub-Clause 12.4. Work that the Employer requires to be omitted from the Contract and performed by others requires that the Employer and the Contractor make a formal supplementary agreement to that effect. A unilateral act by the Employer would be a breach of contract. The Contractor may have already incurred cost in preparing for the execution of

the part of the work to be omitted. As a minimum he would be entitled to payment of the element of profit/risk included in the Accepted Contract Amount.

(e) *'any additional work, plant, materials or services necessary for the Permanent Works'*
The Engineer is entitled to instruct any additional or varied works necessary for the Works to fully function and meet its intended purpose even if these works were not specifically identified in the Contract Documents.

(f) *'changes to the sequence or timing of the execution of the Works'*
It may be assumed that the Engineer will issue instructions under this heading only for the convenience of the Employer. Circumstances for such instructions could include changes in design and the requirements of other contractors. The Contractor will need to carefully review the consequences on his own operations. It is noted that this sub-clause only refers to timing of the execution of the Works, but does not allow the Contractor to alter the timing of completion. Acceleration, if required, is covered in Sub-Clause13.2.

There are occasions when Variations are issued without prior agreement or negotiation. This situation not infrequently arises when additional or revised drawings for construction are issued which include additional or varied works not foreseen in the Contract Documents. The Contractor is to give Notice to the Engineer, who in turn should initiate the Variation procedure without delay in order that the progress of the Works shall not be delayed.

13.2 Value Engineering

'The Contractor may at any time submit to the Engineer a written value engineering proposal at his own cost' but he is under no obligation to do so. The proposal is required to meet one or more of four given criteria:

- it must accelerate the works
- it must reduce the cost to the Employer
- it must improve efficiency or value to the Employer
- it must otherwise be of benefit to the Employer

Any proposal is to be prepared in the manner given in Sub-Clause 13.3 (Variation Procedure) and, if accepted, the Contractor shall be responsible for design (if any).

'The Engineer shall, after receiving the Contractor's proposal, respond by giving a Notice to the Contractor stating his consent or otherwise shall be at the sole discretion of the Employer'. The sub-clause continues *'The Contractor shall not delay any work while awaiting a response'*. It is not reasonable that the Contractor is required to take the risk to start work and incur costs in such circumstances without (at least) the Engineer's consent. It would be a better approach for the Contractor to informally offer the Value Engineering proposal and confirm the same only when the proposal is acceptable to the Employer. If the proposed change results in a reduction in the contract value of this part, the Contractor shall receive 50% of this value (excluding adjustments under Sub-Clauses 13.7 and 13.8).

13.3 Variation Procedure

The same sub-clause numbering and sub-clause headings were used in the First Edition. In this Second Edition, the original sub-clause has been divided into two parts:

Sub-Paragraph 13.3.1 (Variation by Instruction) and
Sub-Paragraph 13.3.2 (Variation by Request for Proposal)

The inference is that Variation by Instruction indicates varied work that will definitely be required, whilst Variation by Request for Proposal indicates varied work which may or may not be finally required.

Sub-Paragraph 13.3.1 states in part '. . . *the Contractor shall proceed with the execution of Variation . . . .*', whereas Sub-Paragraph 13.3.2 states in part '. . . *the Contractor shall not delay any work whilst awaiting a response . . . .*'. In this sub-paragraph, the word 'work' is understood to refer to work already instructed and in progress.

13.3.1 Variation by Instruction

The Engineer may issue a Variation by giving Notice (describing the required change and stating any requirements for recording of Costs) in accordance with Sub-Clause 3.5 (Engineer's Instructions).

The Contractor is required to proceed with the execution of the varied works and within 28 days (or other period as may be agreed) submit to the Engineer detailed particulars including:

(a) description of the varied work performed or to be performed, including details of the resources or methods adopted or to be adopted
(b) a programme for execution, any necessary modifications to the Programme and Time for completion
(c) the Contractor's proposal for adjustment of the Contract Price shall be obtained by valuing the Variation in accordance with Clause 12 (Measurement and Valuation) with supporting particulars (estimated quantities and the cost of EOT if applicable). If there are omissions, the Contractor's proposal may include the amount of any loss of profit or other losses and damages.

The Engineer shall then proceed under Sub-Clause 3.7 (Agreement or Determination) to agree or determine the EOT (if any) and the adjustment to the Contract Price.

The wording of the above requirements leaves much to be desired. Words such as 'Detailed Particulars' and 'proposal' are not defined terms.

- if the Contractor is to provide the Engineer with a 'proposal', he first requires documentation from the Engineer setting out the Engineer's requirements in detail, including applicable specifications, scope of work and an indicative Bill of Quantities. If the appropriate information is already in the possession of the Contractor, he may be able to safely proceed with the execution of the Works (including the Varied Works)
- using the Engineer's documentation, the Contractor can prepare and submit the stated detailed particulars to the Engineer for his agreement or further consideration. These detailed particulars could be divided into two parts:

  * a technical part, including programme and/or programme adjustments
  * a Bill of Quantities with pricing (Not all pricing details may be available in the short term)

The above procedures would facilitate obtaining an agreement under Sub-Paragraph 3.7.1 (Consultation to reach Agreement).

13.3.2 Variation by Request for Proposal

The Engineer may request a proposal by giving a Notice describing the proposed change to the Contractor.

The Contractor is required to respond to the Engineer's request as soon as possible and submit a detailed proposal in accordance with items (a) to (c) of Sub-Paragraph 13.3.1.

Should the Contractor not be able to comply, he shall provide his reasons as described in sub-points (a) to (c) of Sub-Clause 13.1.

If the Contractor submits an acceptable proposal, then the Engineer shall respond by giving a Notice to the Contractor stating his consent or otherwise. The Contractor shall not delay any work whilst awaiting a response. If the Engineer consents to the Contractor's proposal (with or without comments), the Engineer shall instruct the Variation. The Contractor is to submit further particulars if requested by the Engineer.

Should the Engineer not give his consent to the proposal (with or without comments) and the Contractor has incurred Cost preparing his proposal, the Contractor shall be entitled, subject to Sub-Clause 20.2, to payment of such Cost.

There are two unresolved difficulties with the above stated procedures:

- The Contractor will not know if he has submitted an 'acceptable proposal' until the Engineer gives his consent. A time limit for the Engineer to provide his consent should be stated.
- The Contractor is required to start work whilst waiting for a response from the Engineer and will commence incurring Cost (other than the Cost of preparing the proposal). Should the Engineer eventually reject the Contractor's proposal, the Contractor will require reimbursement of that additional Cost together with the overhead Cost and profit.
- If there is doubt concerning payment due to the Contractor, then the Contractor is advised to seek immediate clarification of the Engineer's intentions.

13.4 Provisional Sums

The term 'Provisional Sum' is defined in Sub-Paragraph 1.1.67. A sum is included in the Bill of Quantities in anticipation of executing work which could not be fully described or valued at the Date for Submission of Tenders. The Provisional Sum can only be expended when instructed by the Engineer.

For each Provisional Sum the Engineer may instruct:

(a) Work to be executed for which adjustments to the Contract Price shall be agreed or determined
(b) Plant, Materials, works, or services to be purchased by the Contractor from a Nominated Subcontractor.

The Contractor may be requested to obtain quotations from several subcontractors or suppliers for review by the Engineer, who by Notice shall instruct the Contractor which quotation is acceptable. If the Engineer does not respond within seven days, the Contractor may select the Subcontractor or supplier at his discretion. Appropriate invoices and/or receipts shall be provided to confirm the amounts claimed under this heading.

The Contractor shall be paid:

(1) The actual amounts paid by the Contractor. The cost for delivery on site (or other location specified by the Engineer, which include the purchase price plus transport to site, if not included in the purchase price. If the goods are be imported, the Contractor is entitled to payment of shipping costs, customs charges, including cost of clearing.

For locally obtained goods, the Contractor is entitled to be paid in the currency of expenditure.

Additionally, there may be some ancillary costs to be considered for payment (building work, earthworks, etc.).

(2) A sum for overhead charges and profit calculated as a percentage of the costs described in 1) above. The percentage shall be that stated in the Bill of Quantities Schedule. If there is no such percentage, then the rate stated in the Contract Data shall be applied.

13.5 Daywork

The opening sentence of this sub-clause states: *'If a Daywork Schedule is not included in the Contract, this sub-clause shall not apply'*. However, the subject of Daywork Schedules is not mentioned in the Guidance for the Preparation of Particular Conditions (neither the Contract Data or the Special Provisions). Therefore, it is incumbent on the Employer (or the Engineer on behalf of the Employer) to provide a suitable daywork schedule in the Tender Documents for completion by tenderers.

The United Kingdom offers a different practice which is based on a standard Daywork Schedule. Tenderers are required to offer a percentage adjustment (plus/minus) to the Standard Daywork Schedule. An estimate of the total value of Dayworks forms part of the Contractor's tender offer for tender evaluation purposes.

Frequently site disputes arise concerning the value and applicability of unit rates for labour.

It is accepted practice that the costs of miscellaneous hand-tools, protective clothing, etc. are to be included in the daywork-labour rates. Tenderers should clarify the cost of transporting workers (including idle time) to the place where the daywork will be executed, whether it is to form part of the labour rates or whether the transport and driver will be paid for separately. If there is no separate provision for these costs, then the daywork-labour rates will need to be significantly enhanced to cover these costs.

Frequently site disputes arise concerning the value and applicability of unit rates for labour.

In summary:

- the direct cost of labour includes wages, allowances, leave pay, sick pay, protective clothing, union contribution, pension contributions
- the gross cost of labour is essentially the direct cost plus the Contractor's overhead costs including profit and risk
- the daywork rate for labour is the gross cost as above plus an allowance for travel time, idle time, suspension, and any other factor directly related to the performance of dayworks

The Contract requires the Contractor to provide the Engineer with daily records of resources used in the previous day's work. It frequently happens in international construction that the language of the Contract is the English language, but not all low-level supervisors have adequate language skills to keep records of the standard required. The Contractor's Representative is well advised to ensure that standardised report forms are available for use by the supervisors. Plant and Equipment fleet numbers and employee company-badge numbers can be used to assist the supervisor to overcome language difficulties. A standard form is provided for guidance (Appendix E refers).

It will be noted that this standard form can be utilised in a variety of situations where the Contract requires the Contractor to keep records to support his claims, particularly those where the Contractor is entitled to receive payment of cost.

13.6   Adjustments for Changes in Laws

This sub-clause has been extended and made more explicit than the comparable clause of the First Edition.

'The Contract Price shall be adjusted to take account of any increase or decrease in Costs resulting from change in:'

(a) the Laws of the Country (including changed laws and new laws)
(b) the judicial or official government or implementation of Laws in (a) above
(c) changes in any permit, permission, licence, or approval as described in Sub-Clause 1.13 (Compliance with Laws), which are obtained by the Employer or Contractor
(d) changes in the requirements for any permit, permission, licence, or approval which are obtained by the Contractor made and/or published after the Base Date.

The Employer and/or the Contractor shall give Notice to the Engineer of any adjustments required with detailed particulars.

If the Contractor suffers delay or incurs an increase on Cost resulting any change in Laws, the Contractor shall be entitled to EOT and/or payment for such cost, subject to Sub-Clause 20.2. If there is a decrease in Cost, the Employer shall be entitled to a reduction in the Contract Price, subject to Sub-Clause 20.2.

Tenderers are deemed to have included in their Tender Price the consequences of legislation in force on the Base Date.

To comply with the various time limitations imposed by Sub-Clause 20.2, it is most important that the Contractor promptly obtains information concerning changes in law having an effect on the Contract Price. It should not be overlooked that changes in law can occur in the period between the Base Date and the award of Contract when the Contractor may not have a presence in the country. Should the validity of tenders be extended, there is an increasing likelihood of new legislation affecting the Contract Price being promulgated.

Most changes in legislation require a formal authorisation by Parliament or a state authority of similar status. The authorisation typically appears within a state publication, and it is recommended that the Contractor makes a subscription to such a publication if available. Other sources of information are the major suppliers (e.g. fuels, cement, timber merchants etc.), the trade unions and, not least, the local newspapers. In the preparation of his tender, the Contractor may have already established rates and prices at the Base Date. Evidence of these rates and prices will be required in order to quantify the consequences of any subsequent changes in law.

Finally, it is to be noted that there may be a partial overlap between the entitlements due under this sub-clause and the entitlements due under Sub-Clause 13.8 (Adjustment for Changes in Cost). An appropriate adjustment in required.

13.7 Adjustment for Changes in Cost

This sub-clause opens with the wording *'If Schedule(s) of cost indexation are not included in the Contract, this sub-clause will not apply'*.

Contracts without provisions for 'Adjustment for Changes in Cost' may be acceptable on a short-term contract (say one year duration) but longer-term contracts will have progressively higher uncertainties, and therefore represent an unacceptable risk factor which is likely to deter Contractors from tendering.

The document 'Notes on the Preparation of Special Provisions' (pages 36–37) provides a template for the calculation of adjustment for changes in cost based on the indices method. This method is in use on the most projects using FIDIC documentation.

In most contracts the Contractor is entitled to claim reimbursement of the increase in cost on a limited number of key items, labour, steel, bitumen, cement, etc. In such circumstances, the Contractor will have to include in his Tender Offer an allowance for all other items not specified. The FIDIC document 'Notes on the Preparation of Special Provisions' identified above recognises this problem by proposing a factor labelled *'fixed'* and with a coefficient of 0.10 as provision for *'all items not specified'*.

The Contractor should take care to select indices whose values are readily available from official sources. Other factors which could affect the choice of indices are the currency of payment and the currency in which the Contractor expects to incur most of his expenditure.

Fluctuations in the relative values of currencies are frequently of concern to contractors. Many contracts specify a fixed exchange rate between local and the selected foreign currency. The real exchange rate may during the execution of the Contract become markedly different, often to the Contractor's disadvantage. There is no remedy in the standard General Conditions of Contract for currency fluctuations and the Contractor may find it opportune to take precautionary measures. The advance purchase of foreign currency options is one possible course of action to minimise the effects of any currency fluctuations. Every effort should be made to avoid the use of general indices, such as cost-of-living index which is unlikely to be representative of the changes in cost of material, labour, plant, and equipment used in construction.

Occasionally, it may happen that one element of cost may increase dramatically for reasons that could not have been foreseen at the Base Date. Unless this cost element is included in whole or in part of one of the indices used to calculate the changes in cost, the Contractor is left with a problem, since there is no other clause in the Contract that provides an alternative means of compensation. In equity, it is also correct to state that a dramatic decrease in costs would produce the opposite effect. The only remedy may lie in the laws of the country.

The values of the various indices are prepared by various international agencies and the appropriate Government authorities. It takes time to collect, analyse, and publish the indices. Typically, there is a delay of two to three months before the updated indices are available. In preparing his monthly evaluation of changes in cost (for inclusion in the Monthly Payment Application), the Contractor should use the latest available value of the index as an interim measure to calculate the changes in cost for the months where data is not yet available. These interim values are to be adjusted once the correct values of the indices become available.

14

Clause 14 Price and Payment

The Contract Price is the value of the Works performed in accordance with the Contract at any stage of the Contract. To facilitate payment to the Contractor, the Contract Price is routinely calculated at calendar month intervals (refer to Sub-Clause 14.3 (Application for Interim Payment Certificates)).

14.1   The Contract Price

(a) The Contract Price shall be evaluated in accordance with Sub-Clause 12.3 and is subject to adjustments in accordance with the Contract.

(b) The Contractor shall pay all taxes, duties, and fees. The cost is deemed to be included in the Contract Price.

(c) Quantities set out in the Bill of Quantities are estimated quantities and are not to be taken as the actual and correct quantities to be executed.

(d) Within 28 days of Commencement Date the Contractor is required to provide a price breakdown for each lump sum. Although not so stated, it may be assumed that these price breakdowns are required for using them in the preparation of the monthly Application for Interim Payment. The Contractor's tender office should be requested to supply this detailed information, as it may not be available on Site.

14.2   Advance Payment

The Employer shall make an advance payment as an interest-free loan for mobilisation. The amount of the advance payment (10% of the Contract Price is the norm) shall be paid as stated in the Contract Data (Notes on the Preparation of Special Provisions, Annex E, Page 63 refers).

Guide to the FIDIC Conditions of Contract for Construction: The Red Book 2017, First Edition.
Michael D. Robinson.
© 2023 John Wiley & Sons Ltd. Published 2023 by John Wiley & Sons Ltd.

14.2.1 Advance Payment Guarantee

The Contractor shall provide an Advance Payment Guarantee in the amounts and currencies equal to the advance payment.

In an ideal situation the Advance Payment Guarantee should be provided by the bank used by the Contractor for the receipt of payments of Interim Payments. Since the majority (or all) of the Advance Payment will be paid in foreign currency, the Contractor's principal banker in his own country is likely to be preferred as provider of the Advanced Payment Guarantee.

The use of smaller, less capitalised banks is not ideal from the Employer's point of view, particularly if the Advance Payment Guarantee is frequently extended by the Contractor. This is an indicator of financial stress in the Contractor's affairs.

The amount of the Advance Payment Guarantee shall progressively be reduced by the amount repaid by the corresponding deduction in each Interim Payment Certificate.

Should the amount of the advance payment not be fully repaid by the date 28 days before the expiry date, then the Contractor shall extend the validity of the guarantee until the advance payment has been repaid. The Contractor shall provide appropriate evidence of this extension, otherwise the Employer shall be entitled to claim under the guarantee the outstanding amounts.

14.2.2 Advance Payment Certificate

There are two steps to be accomplished in the process leading to the Employer making the Advance Payment at the earliest opportunity.

Firstly, the Employer must be in the possession of both the Performance Security and the Advance Payment Guarantee. The Contractor will be well aware of the importance of fulfilling this obligation without delay and will have made preliminary arrangement for the provision of these documents during the pre-tender period.

Secondly, the Contractor's application for the Advance Payment is to be provided and agreed with the Engineer (the format of which having pre-agreed with the Engineer). Although this application is not a complicated document, it may take some time to agree the format, particularly if the Employer and/or loan agencies also have to be consulted.

It is recommended that this advance payment application be given Index 0, as it does not relate to any particular time period. Subsequent applications can then be numbered 1, 2, 3 . . . . representing the value of work performed in the subsequent months of execution of the Contract.

It is to be noted that the advance payment is specifically intended to finance the Contractor's mobilisation costs. Although not so stated in the General Conditions of Contract, it is possible that the Contract Documents may contain provision for the Contractor to provide evidence that he has spent the advance payment specifically for this purpose.

14.2.3 Repayment of the Advance Payment

The advance payment shall be repaid by percentage deductions from Interim Payment Certificates.

Deductions will commence when the running total of IPCs exceeds 10% of the Accepted Contract Amount (less Provisional Sums). Thereafter deductions will be made at a rate of 25% from each of the subsequent IPCs (excluding the advance payment itself and any release of retention money) until the total of the advance payment has been repaid).

Typically, the advance payment will have been repaid when approximately 75–80% of the Works have been completed. However, if the total of the advance payment has not been repaid before the issue of the Taking-Over Certificate, then the whole of the balance still outstanding shall immediately become due and payable by the Contractor. The Employer may use the Advance Payment Guarantee to make good the deficit. Alternatively, the Contractor may agree that any return of retention money to which he may be entitled can be used to pay the outstanding amount of Advance Payment.

14.3 Application for Interim Payment Certificates

The First Edition required that a copy of the Progress Report be provided as part of the supporting documents for any Application for Interim Payment Certificates. This requirement has been deleted in the Second Edition.

The Contractor is required to submit a Statement to the Engineer after the end of each month, showing in detail the amounts to which the Contractor considers himself to be entitled, together with all supporting documents including the report described in Sub-Clause 4.21 'Progress Reports'. The above extract quite clearly entitles the Contractor to include in his application the amounts to which he, the Contractor, considers himself entitled. The preparation of the corresponding Interim Payment Certificate for presentation by the Engineer to the Employer is the duty and responsibility of the Engineer and does not contractually involve the Contractor. Despite this wording, invariably the Engineer and/or the Employer require the Contractor to modify his application to conform to the certification of the Engineer. In some jurisdictions the Employer requires the Contractor to sign the Engineer's Interim Payment Certificate in order to comply with the Employer's internal accountancy procedures. This comment particularly applies where the Employer is a government agency or parastatal company. This procedure is not provided for in the Conditions of Contract and Employers generally do not consider it necessary to specify their precise requirements by including modified clauses in the Particular Conditions of Contract. Sub-Clause 14.3 (c) clearly states in part (the Contractor shall show) *'in detail the amount [he] considers himself to be entitled . . .'*. Doubt may arise that, by signing, the Contractor accepts the Engineer's figures to represent his full entitlement.

Regrettably, this non-contractual process often can obscure and diminish the very real concerns that the Contractor may have in respect of unpaid items and disputes. Under the circumstances described above, it is recommended that the Contractor prepare a further listing of claims and other unresolved issues and their valuations not included in the Engineer's Interim Payment Certificate. This letter is to be sent to the Engineer, copy to the Employer, stating that 'the listing is a summary of all claims and unresolved issues and their valuations as at (date) which are not included in the Engineer's Interim Payment Certificate No . . . . for the month of . . . '.

This listing should be prepared in sufficient detail to comply with the requirements of Sub-Clause 4.20 (f), as this will avoid duplication of effort.

The Contractor's Statement shall include:

(a) the estimated contract value of the Works produced up to the end of the month. This is essentially a measurement of the items included in the Bill of Quantities. Most of the measurement will relate to physical work, but some administrative items are likely to require evaluation.

(b) *'any amounts to be added or deducted for changes in legislation (Sub-Clause 13.7) and changes in cost (Sub-Clause 13.8)'*

(c) Any amount to be deducted for retention up to the limit stated in the Contract Data (Page 5, Limit of Retention). A partial Taking Over of the Works would entitle the Contractor to a partial return of retention (refer to Sub-Clause 14.9).

(d) *'any amounts to be added and deducted for the advance payment and repayments'*

(e) *'any amounts to be added and deducted for Plant and Materials conforming to the requirements of Sub-Clause14.5'*

(f) *'any other additions or deductions due under the Contract'* including claims, variations, dayworks

(g) previous certified amounts shall be deducted.

It is strongly recommended that the Contractor does not wait until the last day of the month before commencing the preparation of this Statement. Many items can be measured as the Works proceed: field surveys can be made; administrative payment items such as invoices can be prepared in advance etc.

The Contractor could consider commencement of the measurement by the 25th day of each month. Draft copies of the sections of the measurement could be submitted to the Engineer's staff in a progressive manner for checking, the last few items being quickly evaluated at the month's end.

Particularly in respect of uncomplicated projects, the Engineer's measurement staff should not require the full allowed period of 28 days to agree an interim measurement for the purposes of producing a pre-agreed Statement and Interim payment Certificate. Both the Engineer's and Contractor's staff entrusted with this task should have a clear understanding of the significance of the word 'interim' as used in this sub-clause. However, it is an unfortunate fact that in many countries even the interim measurements must be precise and wholly accurate even though they will be out of date before they are finalised.

14.4   Schedule of Payments

The Contract may provide a Schedule of Payment as an alternative to the measurement process described in Sub-Clause14.3 (Application for Interim Payment).

The use of a Schedule of Payments system is best suited to a contract where the work is substantially repetitive with minimal involvement of nominated subcontractors. At the Tender Date nominated subcontractors may only be identified by the inclusion of a Provisional Sum in the Bill of Quantities. Payments to nominated subcontractors would therefore require their own separate Schedule of Payments to be prepared once the Subcontract is finalised.

A Schedule of Payments may require adjustment if:

(a) the Contractor's proposed Schedule of Payments requires adjustment if it is shown to be inaccurate. Considering that the Engineer (Design) will have produced both an initial programme of work and cost estimates for use by the Employer in the preparation of the Tender documents, the accuracy of the Contractor's own proposal could be assessed reasonably well, avoiding the formal time-consuming negotiation process envisaged in the penultimate paragraph of this sub-clause.
(b) There may be programme slippages (whatever the cause) which will delay the Works and the Contractor's earnings.

The last clause envisages a contract arrangement wherein the Contractor provides the Engineer with estimates of payments to become due in the next three months. Revised estimates are to be submitted at three monthly intervals. Over/under-estimates may be expected. This arrangement is best limited to short term contracts with repetitive work dominant.

14.5 Plant and Materials intended for the Works

This sub-clause provides for the Interim Payment Certificates to include amounts for Plant and Materials that have been delivered to Site for incorporation in the Permanent Works. Payments under this sub-clause may be limited to specific items of Plants and Materials identified in the Contract Data (Page 5) which are categorised as:

(1) Plant and Materials for payment when shipped or
(2) Plant and Materials when delivered to the Site

These descriptions are repeated in the text of Sub-Clause 14.5 and additional wording added requiring that these interim amounts be deducted when the Plant and Materials are included in the Permanent Works, when payment in full becomes due.

It is not practical for every minor item to be considered, not least because the administrative costs would probably exceed the potential benefits to the Contractor.

In respect of Plant and Materials to be shipped to the Site from another country, two possibilities exist:

(a) Payment is permitted once the Plant and Materials are shipped but not yet delivered to Site. A bank guarantee similar to that required for Advance Payment (Sub-Clause 14.2) has to be provided. This guarantee must be valid until the Plant and Materials are on site and safely stored and protected.
(b) Payment is permitted once the Plant and Materials are on site and safely stored and protected.

The Contractor will be required to produce documentary evidence of costs and evidence demonstrating that the Plant and Materials are in conformity with the Contract. The latter requirement may demand testing and/or inspection certification on or off the Site.

More commonly the above requirements will permit payment for other permanent materials (e.g. cement, bitumen, reinforcement etc.) purchased locally and delivered to Site. In addition, payment may be due for permanent materials manufactured on site by the Contractor. Typically, this would include crushed stone, precast items and similar. Again, evidence of cost and evidence of conformity with the contract specifications are required. The cost of materials manufactured by the Contractor will have to be negotiated with the Engineer.

Materials on Site will be progressively consumed as the Works proceed but will be replaced as new Materials are manufactured. Consequently, the value of Materials on Site will rise or fall when compared with the value at the end of the previous month. Eventually the value will become zero as the operations incorporating the Materials on Site are completed.

Plant on Site will eventually be installed as the Works proceed and will become included in the bill items forming part of the IPC.

In all cases the amount to be certified by the Engineer is the equivalent of 80% of the agreed amounts. The currency split will be the same as provided in the relevant Bill of Quantities item.

Proof of ownership of Plant and Materials can be problematic, particularly if the subcontractors and sub-suppliers are involved. Sub-Clause 7.7 recognises that there may be problems: *Each item of Plant and Materials shall, to the extent consistent with the Laws of the Country, become the property of the Employer . . . '.* This is potentially a complex issue, and the Employer may require additional guarantees from the Contractor in order to protect his interests.

14.6   Issue of IPC

It is a pre-condition that no amount will be certified or paid to the Contractor until the Contractor has provided the Employer with the Performance Security (Sub-Paragraph 4.2.1) and the Contractor has appointed the Contractor's Representative (Sub-Clause 4.3). It is noted that these pre-conditions do not apply to the Advance Payment (Sub-Clause 14.2).

14.6.1   The IPC

The Engineer shall within 28 days after receiving a Statement issue an IPC to the Employer with a copy to the Contractor stating the amount which the Engineer fairly considers to be due, including any additions and/or deductions which have become due (Sub-Clause 3.7 (Agreement or Determination)).

14.6.2   Withholding (Amounts in) an IPC

The Engineer may decline to issue an IPC should the amount of the IPC be less than the minimum amount (refer Contract Data Page 5). This situation may arise in the first month(s) of the Contract in which case the Contractor is unlikely to make an Application for Interim Payment (Sub-Clause 14.3). A similar situation may arise closer to the date by which the Engineer anticipates the Taking-Over Certificate will be issued. Whilst not obliged to do, the Employer may decide to allow the Engineer to issue the IPC.

Once the Taking-Over Certificate issued, no such restriction exists. Retention Money is to be repaid to the Contractor together with any residual amounts due for minor outstanding work and claim settlements. Following discussions with the Engineer, an additional IPC may be required.

14.6.3 Correction or Modification

The Engineer may in any Payment Certificate make any correction or modification that should properly have been made in a previous Payment Certificate.

Should the Contractor disagree with corrections or modifications made by the Engineer, he shall make clear his disagreement by including the affected items in a further Payment Certificate. Thereafter, if the Engineer does not agree to modify this Payment Certificate, then the Contractor may give a Notice referring the matter to the Engineer for Agreement or Determination (Sub-Clause 3.7 refers).

Note:
The time limit for agreement commences on the date of the Notice.

14.7 Payment

The Employer shall pay to the Contractor

(a) the amount in each Advance Payment Certificate within the period stated in the Contract Data. If the number of days is not stated in the Contract Data, it shall be taken as 21 days.
(b) the amount in each IPC (Sub-Clause 14.6) within the period stated in the Contract Data. If the number of days is not stated in the Contract Data, it shall be taken as 56 days.
(c) the amount due on the issue of the FPC (Sub-Clause 14.13) within the period stated in the Contract Data, which, if not stated, shall be 28 days after the Employer receives the IPC.
(d) the amount certified in the FDC with the period stated in the Contract Data which, if not stated, shall be 56 days after the Employer receives the FPC.

14.8 Delayed Payment

Employers may be reliant on project financing from external funding agencies with which to make payments to the Contractor. Occasionally it may happen that the Employer experiences administrative difficulties in the timely provision of those funds and is unable to pay the Contractor as required. Funding agencies are reluctant to allow their funding to be used for payment of financing charges to the Contractor.

'If the Contractor does not receive payment in accordance with Sub-Clause 14.7 above, he is entitled to payment of financial charges at an annual rate of three percent above:

- *the average bank short-term lending rate to prime borrowers for the currency of payment at the place of payment or*
- *where no such rate exists at that place, the same rate in the currency of the country of payment or*
- *in the absence of such rate at a place, the appropriate rate fixed by law of the country of the currency of payment.*

The period of delay is deemed to commence on the expiry of the time for payment (Sub-Clause 14.7)'.

The Contractor is not required to submit a Statement or formal Notice.

14.9   Release of Retention Money

1. Taking Over of the Whole of the Works
   - (a) Upon the issue of a Taking-Over Certificate for the whole of the Works, the Contractor shall include the first half of the Retention in a Statement. In the next IPC, after the Engineer receives such Statement, he shall certify the release of the first half of the Retention Money.
   - (b) After the latest of the expiry dates of the DNP the Contractor shall include the second half of the Retention Money in a Statement. In the next IPC the Engineer shall certify the second half of the Retention Money.
2. Taking Over of Sections of the Work
   - (a) Upon the issue of a Taking-Over Certificate for a Section of the Works, the Contractor shall include the first part of the Retention Money in a Statement. In the next IPC the Engineer shall certify the release of the first half of the Retention Money.
   - (b) After the latest of the expiry dates of DNP, the Contractor shall include the second half of the Retention Money in a Statement. In the next IPC the Engineer shall certify the release of the second half of the Retention Money.
   - (c) The relevant percentages for each section shall be given in the Contract Data. If no percentage values are given in the Contract Data, then no percentage half for Taking Over of Sections shall be made.
3. Taking Over of Parts of the Work
   Taking Over Parts of the Works is governed by Sub-Clause 10.2 (Taking Over Parts) and does not provide for any return of Retention but does provide for a reduction of Delay Damages.
4. Cost of Remaining Work
   When certifying any release of Retention Money, the Engineer is entitled to withhold the estimated cost of any work remaining to be executed.

14.10   Statement on Completion

'Within 84 days after receiving the Date of Completion for (the whole of) the Works, the Contractor shall submit to the Engineer a Statement at Completion showing'

(a) *'the value of all works done in accordance with the Contract at the date of the Taking-Over Certificate'.* This would include the measured works, the value of varied works performed, the value of accepted claims, as well as any amounts due under Sub-Clauses 13.7 (Changes in Legislation) and 13.8 (Changes in Cost).
(b) *'any further sums which the Contractor considers to be due'.* This would include an estimate of the value of unresolved claims including ongoing measurement disputes.
(c) *'an estimate of any other amounts which the Contractor considers may become due to him under the Contract'.* This would include the value of ongoing events and claims after the date of issue of the Date of Completion of the Works. The value of any outstanding work should also be included. It is not for the Contractor to anticipate any future additional work which might be instructed by the Engineer.
The Statement on Completion shall include:
  - claims for which the Contractor has submitted a Notice under Sub-Clause 20.2 (Claims for Payment and EOT)
  - any matter referred to the DAAB under Sub-Clause 21.4 (Obtaining DAAB Decision)
  - any matter for which a NOD has been given under Sub-Clause 21.4 (Obtaining a DAAB Decision)

It is important that the Contractor evaluates the Statement at Completion accurately. It represents the maximum revenue that could be due to him and equally indicates to the Employer the likely maximum amount of his financing commitment.

14.11   Final Statement

This Final Statement is effectively an update of the Statement at Completion (Sub-Clause 14.10 refers) and is submitted by the Contractor once the Performance Certificate is issued by the Engineer. The issue of the Performance Certificate is an affirmation that the Contractor's obligations under the Contract have been completed (Ref. Sub-Clause 11.9 (Performance Certificate)).

14.11.1   Draft Final Statement

Within 56 days of the issue of the Performance Certificate, the Contractor shall submit to the Engineer a draft final statement which shall:

- be in the same form as the IP statements submitted under Sub-Clause 14.3 (Application for Interim Payment)

- show in detail
  - (i) the value of all work done in accordance with the Contract
    At this stage of the Contract the final measurement of the Works should be complete excepting for very few outstanding items. In effect it is a minor updating of the penultimate IPC.
  - (ii) any further sums which the Contractor considers to be due at the date of the issue of the Performance Certificate
    During the Defects Notification Period it is likely that the Contractor will have executed a number of minor works (not being repairs of defects) for which the Contractor is entitled payment. It is likely that the sums involved are minor. Also, it is possible that some administrative items (not previously quantifiable) are to be added.
  - (iii) an estimate of any other amounts which the Contractor has or will become due after the issue of the Performance Certificate
    This estimate will be the total amount of unresolved Contractor claims including those which may have been referred as a Dispute to the DAAB. The Engineer will be aware of these claims and their value. However, should the Engineer disagree or be unable to verify any part of the Contractor's Draft Final Statement, then by Notice to the Contractor he may request clarification or further details.

14.11.2   Agreed Final Statement

If there are no amounts under sub-section (iii) of Sub-Paragraph 14.11.1, then the Contractor shall prepare and submit to the Engineer the Final Statement. At this stage of the Contract, it may be assumed that the Employer and the Contractor would agree to discuss outstanding issues and reach an amicable solution.

However, if outstanding issues are not resolved, then the Contractor shall prepare and submit to the Engineer a Partially Agreed Final Statement, identifying separately the estimated amounts and the disagreed amounts.

14.12   Discharge

When submitting the Final Statement (or the Partially Agreed Final Statement) the Contractor shall provide written confirmation (a letter of 'discharge') that the total of the Statement is a final and full settlement of all moneys due to the Contractor under or in connection with the Contract, excluding any amount that may become due in respect of any Dispute for which a DAAB proceeding or arbitration is in progress.

The letter of discharge should also state that it becomes effective upon the Contractor receiving the full amount certified in the Final Payment Certificate (FPC) and the Performance Security has been returned.

In consideration of the importance of this letter of discharge, it is recommended that the drafting of the letter be provided by the Contractor's legal advisors. The preparation and submittal of the letter of discharge should not be unduly delayed, as the discharge is stated to be effective once the Final IPC has been paid and the Performance Security returned. The above does not affect either party's liability or entitlement in respect of any dispute which is in progress (ref. Clause 21 (Disputes and Arbitration)).

14.13 Issue of Final Payment Certificate (FPC)

Within 28 days after receiving the Final Statement (or the Partially Agreed Final Statement) the Engineer shall issue the Final Payment Certificate (FPC) which shall:

(a) state the amount which the Engineer considers is finally due to the Contractor after giving credit to the Employer for all amounts already paid by him
(b) give due allowance for other debits and credits due between the parties.

If the Contractor has not provided a Draft Final Statement within the time specified under Sub-Paragraph 14.11.1 (Draft Final Statement), the Engineer shall again request the Contractor to do so. Should the Contractor not respond within a further period of 28 days, the Engineer shall issue the FPC for the amount the Engineer considers is due.

However, if the Contractor has submitted a Partially Agreed Final Statement or, alternatively, if the Contractor has submitted a Draft Final Statement (Sub-Paragraph 14.11.1 refers) which corresponds to the Partially Agreed Final Statement, the Engineer shall proceed in accordance with Sub-Clause 14.6 (Issue of IPC), to issue an IPC.

Contractors should be aware that a failure to cooperate with the Engineer to follow the specified procedures leading to the issue of a FPC will not prevent the Engineer fulfilling his contractual duties.

14.14 Cessation of Employer's Liability

'The Employer shall not be liable to the Contractor for any matter or thing in connection with the Contract, excepting to the extent that the Contractor shall have included an amount expressedly for it

(a) *in the Final Statement (or Partially Agreed Statement)*
(b) *except for matters arising after the issue of the Taking-Over Certificate (refer Sub-Clause 14.10) in the Statement described in Sub-Clause 14.10'.*

The contents of this Sub-Clause emphasize the obligation of the Contractor to check the Final Statement very carefully to ensure that he has included for all claims and payment issues (refer Sub-Clause 14.10).

This sub-clause does not limit any liability of the Employer in respect of his indemnification obligations or liability arising from fraud, deliberate default, or reckless misconduct.

14.15 Currencies of Payment

(a) The general practice is that the Contract Price is quoted in the local currency with the proviso that a fixed percentage shall be paid in foreign currency at a fixed exchange rate for the duration of the Contract.

The Advance Payment (refer Sub-Clause 14.2) will be paid in the same proportions unless otherwise stated or agreed.

(i) Any taxes, duties and fees which are to be paid by the Contractor and are known to him at the date of tender will likely have been allowed for in the calculation of general overheads. However, if new taxes are introduced or existing taxes increased, then the Contractor would be entitled to make a formal claim of the additional expenditure.

(ii) The calculation of the amounts due for works described as Provisional Sums (refer Sub-Clause 13.4) is dependent on the type of work to be carried out. Foreign currency may be expended for imported industrial goods, whilst site activities may be valued at existing bill rates. Consequently, payment for Provisional Sums may have a unique local/foreign currency split.

It is recommended that the Contractor ensures that all quotations, invoices, or services from foreign suppliers are priced solely in the local currency or the foreign currency of the Contract. This procedure will avoid any time-wasting verifying exchange rates between differing foreign currencies.

Adjustments for Changes in Law (ref. Sub-Clause 13.6) refer to local changes which might be promulgated in a State Gazette or similar official document. These changes may affect the price of petroleum products, cement prices, labour rates and conditions. The amount of any increase (or decrease) in cost would be evaluated in local currency only.

(iii) The following items shall be paid (or repaid) in the same currency proportions as given in the Contract Data Item 14.15 (a)(i)
- the value of work executed
- the amounts to be deducted for Retention and the amounts of Retention returned
- the amount of advance made and subsequent repayment

(b) Whenever an adjustment (Value Engineering or Variation) is agreed, the applicable currencies shall be specified.

(c) Payment of Delay Damages shall be made in the proportions specified in the Contract Data (Page 6, Items 14/15c)

(d) Other payments to the Employer by the Contractor shall be made in the currency specified in the Contract (where not otherwise identified).

(e) If in any given period the Contractor has a net debit in favour of the Employer in one of the contract currencies, the Employer may recover the balance in another currency.

(f) Exchange rates not stated in the Contract Data shall, if required, be those prevailing in the Base Data and published by the central bank of the country.

The Contractor should give consideration as to how he might protect himself against excessive exchange rate fluctuations during the execution of the Contract.

15

Clause 15 Termination by the Employer

Note:
This Clause deals with two differing causes which may give rise to a termination of the Contract:

- Sub-Clause 15.2 (et al.) deals with termination of the Contract because of the Contractor's Default.
- Sub-Clause 15.5 (et al.) deals with termination of the Contract for the Employer's convenience.

In addition, termination of the Contract by the Contractor due to default of the Employer is described in Clause 16, and termination of the Contract as a consequence of an Exceptional Event is described in Sub-Clause 18.5 (Optional Termination). The obligations and entitlements of the Parties vary considerably, according to the reasons causing the termination.

15.1 Notice to Correct

The opening sentence of this sub-clause states:
'If the Contractor fails to carry out any obligation under the Contract, the Engineer may, by giving a Notice to the Contractor, require the Contractor to make good the failure. . . .'
'Obligation' can be broadly interpreted as a failure. Failure to correct substandard work, failure to provide samples, and more importantly failure to proceed in accordance with the Programme, failure to train employees of the Employer in good time etc. The listing is potentially endless.

There is a danger that the over-use of this sub-clause with respect to the many minor failures that will regularly occur on a given project will be counterproductive. In particular the field staff of the Engineer and Contractor should be encouraged to work together to minimise or eliminate the possibility of failures occurring and if they do occur, ensure they are corrected without delay. There are many opportunities for the need for correction of a fault to be discussed on the work site and in site meetings. More significant failures requiring the attention of senior management should always be discussed before a Notice is issued by the Engineer. It must not be assumed that all failures derive from a fault of the

Guide to the FIDIC Conditions of Contract for Construction: The Red Book 2017, First Edition.
Michael D. Robinson.
© 2023 John Wiley & Sons Ltd. Published 2023 by John Wiley & Sons Ltd.

Contractor. The overall intent should be to identify and execute the correct solution with minimum delay.

Consequently, this Sub-Clause 15.1 should only be actioned when other solutions are exhausted. The remaining portion of this sub-clause specifies the intended contents of the Notice to Correct.

15.2 Termination for Contractor's Default

15.2.1 Notice

'The Employer shall be entitled to give a . . . Notice. . . of the Employer's <u>intention to terminate the Contract</u>. . . or in the case of sub-paragraph (f), (g) or (h) below a <u>Notice of Termination</u>' *(underscoring added).*

The implication is that should the Contractor take the appropriate corrective action in respect of items (a) to (e), the Employer may reconsider whether to terminate the Contract. However, the matters referred to in items (f) to (h) cannot be corrected and the Contract is to be terminated if the Contractor

(a) fails to comply with:
   (i)   a Notice to Correct
   (ii) a binding agreement under Sub-Clause 3.7
   (iii) a decision of the DAAB under Sub-Clause 21.4
   and such failure constitutes a material breach of the Contractor's obligations under the Contract.
(b) abandons the Works
(c) without release fails to proceed with the Works in accordance with Clause 8 (Commencement, Delays and Suspension) or
   the Contractor's failure to comply with Clause 8 entitles the Employer to Delay Damages in excess of the maximum stated in the Contract Data
(d) fails to comply with a Notice of Rejection (Sub-Clause 7.5)
(e) fails to comply with Sub-Clause 4.2 (Performance Security). The Performance Security is a key financial security specifically intended to ensure completion of the Works.
(f) subcontracts the whole or part of the Works without permission (ref. Sub-Clause 1.7 (Assignment))
   It is a not uncommon ply that a Contractor in financial difficulties will engage sub-contract labour without the knowledge or permission of the Engineer. Almost inevitably the subcontractor will not be paid and will leave the site without notice.
(g) becomes bankrupt or insolvent etc.
   It frequently happens that the contract revenues projected by the Contractor are being achieved. However, other contracts or financial issues are draining the Contractor's resources to an extent that his whole business becomes insolvent leading to bankruptcy.
   or if the Contractor is a JV

(i) any of the above matters apply to a member of the JV
(ii) the other members of the JV do not promptly confirm to the Employer the retiring member's obligations under the Contract shall be fulfilled in accordance with the Contract.

In the event of one member of a JV becoming bankrupt or otherwise unable to meet their obligations, the standard practice is that the remaining members of the JV will proportionately take up the share of the departing member. For legal reasons the departing member will retain a minimum share (e.g. 0.01%) but without right of further participation in the affairs of the JV. A second possibility is that a replacement partner will be found to take up the share of the departing member.

A JV is itself a registered company and is obliged to fully inform the Employer of these proposals and obtain his permission.

(h) the JV has engaged in fraudulent, collusive, or coercive practices.

15.2.2 Termination

Unless the Contractor remedies the matter described in Sub-Paragraph 15.2.1 (Notice) within 14 days or receiving the Notice, the Employer may by giving a second Notice to the Contractor immediately terminate the Contract. The date of termination shall be the date the Contractor receives this second Notice.

However, in the case of Sub-Paragraph 15.2.1 items (f), (g) or (h) the Employer may by giving a Notice immediately terminate the Contract. The date of termination shall be the date the Contractor receives the Notice.

The Notice of Termination should be delivered by hand or courier to the Contractor's registered address and a signed and dated receipt obtained.

As soon as the Termination is announced, the site must be secured by the Employer in order to prevent the removal of any item that is not personal.

15.2.3 After Termination

After termination of the Contract the Contractor shall:

(a) comply immediately with the instructions of the Employer
   – for the assignment of any subcontract
   – for the protection of life or property
(b) deliver to the Engineer
   (i) any goods required by the Employer
   (ii) all Contractor Documents and
   (iii) all other design documents made by or for the Contractor for design of the Permanent Works for which he is responsible
(c) leave the Site.

The above requirements do not reflect the reality of a Contract that is terminated for several reasons, including:

1. It will be very difficult to persuade the Contractor's employees to remain on Site, not least because they will be concerned that their salaries/wages will not be paid. Any expatriate staff are likely to request repatriation to their own countries. The Contractor's staff are to be allowed to remove their personal belongings without hindrance.
2. It is important that the Site is immediately secured by the Employer's own personnel. The Employer should prepare for this eventuality in advance of the issuing of the Notice of termination.
3. A full inventory of the Site is to be taken, including:
   - All Contractor's Goods (refer also to Sub-Clause 4.17)
   - Subcontractors' trucks, personnel carriers (Some of these items may be owned by third parties who will request release of their property.)
   - All suppliers' materials delivered to Site and paid for the Employer in an IPC. Suppliers' materials delivered to Site but not yet paid for should not be removed. However, if it is intended that a replacement contractor shall use the materials, the Employer should pay for them.
4. The condition of the Site must be accurately recorded. The quantities of work properly performed at the Date of Termination are also to be recorded. This will enable a bill of quantities required for a replacement contract to be quickly drawn up.

15.2.4 Completion of the Works

After termination the Employer may complete the Works himself or arrange for other entities to do so. The Employer or other entities may complete the Works using any Goods and Contractor's Documents to complete the Works. (Note: It is presumed that this procedure is permitted by the applicable law or if the Contractor is entitled to payment for use of the Contractor's property). Following completion of the Works, the Contractor's Equipment and Goods will be released to him at Site.

However, if the Contractor is indebted to the Employer, the Employer may sell off items of Equipment to recover this debt. There is no mention of the technical condition of the released Equipment after use by others. The Official Receiver of a bankrupt Contractor may have no further interest in this matter.

15.3 Valuation after Termination for Contractor's Default

'Following termination of the Contract the Engineer shall proceed to agree or determine the value of the Permanent Works, Goods and Contract Documents and any other sums due to the Contractor for work executed in accordance with the Contract. This valuation shall not include the value of any Contract Documents, Materials, Plant and Permanent Works to the extent they do not comply with the Contract'.

Sub-Clauses 15.3 (Valuation After Termination for Contractor's Default) and Sub-Clause 15.4 (Payment After Termination for Contractor's Default) describe how the Employer's entitlement to recovery of additional costs, losses and damages and Delay Damages shall be evaluated. Inevitably, this process will be a lengthy and complex process. Sub-Clause 15.4 describes how the elements of Employer's Costs shall be evaluated, which notably includes provision for payment of Delay Damages.

If the Contractor is bankrupt and without appropriate staff, it is not clear who will represent the Contractor in these matters.

15.4   Payment after Termination for Contractor's Default

The Employer may withhold payment to the Contractor of the amounts agreed or determined under Sub-Clause 15.3 until the following have been established:

(a) the additional costs of executing the (remaining) Works
(b) any losses and damages suffered by the Employer in completing the Works. (Prolongation costs in respect of Employer's personnel may be significant.)
(c) Delay Damages

15.5   Termination for Employer's Convenience

In the First Edition this sub-clause states in part
   'The Employer shall not terminate the Contract under this sub-clause in order to execute the Works himself or to arrange for the Works to be executed by another contractor'.
   In this Second Edition this sub-clause is retitled 'Termination for the Employer's Convenience'. The quotation given above has been deleted and replaced by the following:
   'Unless and until the Contractor has received payment of the amount due under Sub-Clause 15.6 (Valuation after Termination for Employer's Convenience), the Employer shall not execute (any part of) the Works or arrange for (any part of) the Works to be executed by any other entities'.
   Whereas in the First Edition the consequences of Employer's termination for his own convenience were not necessarily favourable to the Employer as effectively he could not complete the Works, the wording of the Second Edition requires only that 'appropriate payment' is to be made to the Contractor of the amounts due under Sub-Clause 15.6 (Value after Termination for Employer's Convenience).
   The Employer shall be entitled to terminate the Contract at any time for the Employer's convenience by giving a Notice referencing this sub-clause. Thereafter the Employer shall immediately:

(a) have no right to use any of the Contractor's Documents unless the Contractor has received payment
(b) have no right to allow use of any Contractor's Equipment etc.
(c) shall return the Performance Security to the Contractor.

   Termination takes effect 28 days after the date of the Contractor receiving this notice or the Employer has returned the Performance Security.

Unless and until the Contractor has received any amounts due under Sub-Clause 15.6, the Employer shall not execute (or arrange for others to execute) any part of the Works.

After termination the Contractor shall proceed in accordance with Sub-Clause 16.3.

15.6 Valuation after Termination for Employer's Convenience

Following termination under Sub-Clause 15.5 the Contractor shall submit:

- the value of work done, including matters described in sub-paragraphs (a) to (e) of Sub-Clause 18.5 (Optional Termination) and
- any additions or deductions and the balance due by reference to sub-paragraphs (a) and (b) of Sub-Clause 14.13 (Issue of FPC).

The Engineer shall agree or determine the amounts due to the Contractor and shall issue a Payment Certificate without the need for the Contractor to submit a Statement.

The agreement of any amounts due will be a lengthy process with particular reference to the valuation of the Contractor's infrastructure.

15.7 Payment after Termination for Employer's Convenience

The Employer will pay the Contractor the amount certified under Sub-Clause 15.6 within 112 days after the Engineer receives the Contractor's submission under that sub-clause.

16

Clause 16 Suspension and Termination by the Contractor

16.1   Suspension by the Contractor

This sub-clause identifies four potential breaches of the Employer's contractual obligations which would entitle the Contractor to suspend the Works:

(a) the Engineer fails to certify in accordance with Sub-Clause 14.6 (Issue of IPC)
(b) the Employer fails to provide evidence of his financial arrangements in the Contract Data (Sub-Clause 2.4 (Employer's Financial Arrangements))
(c) the Employer fails to comply with Sub-Clause 14.7 (Payment)
(d) the Employer fails to comply with binding agreements (Sub-Clause 3.7 (Agreement or Determination)) or a decision of the DAAB (Sub-Clause 21.4 (Obtaining DAAB's Decision)).

All four headings relate to financial events which would delay or deprive the Contractor of payments rightfully due under the terms of the Contract. Disruptions to the Contractor's cash-flow can have unintended consequences to the routine progress of the Works on resumption of the Works.

Public authority employers with financing provided by international financing agencies are less likely to interrupt the Contractor's cash-flow. However, contracts to be executed using private funding may cause the Contractor to request strong financial guarantees from the investor.

In the event of a breach of contract by the Employer as described above, the Contractor is entitled to give the Employer a Notice of Suspension of the Works. The Notice of Suspension becomes effective not less than 21 days after the giving of the Notice. The suspension continues until the Employer rectifies the default, after which normal working shall recommence as soon as possible.

The Contractor is entitled to payment of financial charges as a consequence of the delay in payment (Sub-Clause 14.8 (Delayed Payment)).

The Contractor is also entitled to reimbursement of any additional costs plus profit incurred because of the suspension of the Works and to EOT (Sub-Clause 20.2 (Claims for Payment and/or EOT)). Additional costs arising from a suspension of the Works are likely to far exceed the additional financing costs.

Guide to the FIDIC Conditions of Contract for Construction: The Red Book 2017, First Edition.
Michael D. Robinson.
© 2023 John Wiley & Sons Ltd. Published 2023 by John Wiley & Sons Ltd.

16.2 Termination by Contractor

'Termination of the Contract under this Clause shall not prejudice any other rights of the Contractor, under the Contract or otherwise'.

16.2.1 Notice

This sub-clause identifies 11 causes which entitle the Contractor to either

A   give a Notice to the Employer which shall state that it is given under this sub-clause with an intention to terminate the Contract. This wording implies that the Contractor intends to terminate the Contract unless the Employer corrects the defaults within a reasonable time
   or
B   give a Notice to the Employer which states that the Contractor is terminating the Contract.

Items falling under Category A above:
(a) the Employer does not provide evidence of financial arrangements
(b) the Engineer fails to issue IPC
(c) the Contractor does not receive amount due under any Payment Certificate within 42 days after expiry of time for payment
(d) the Employer fails to comply with
    (i) a binding agreement made under Sub-Clause 3.7
    (ii) a decision of the DAAB under Sub-Clause 21.4
(e) the Employer substantially fails to perform (a material breach of the Employer's obligations under the Contract)
(f) the Contractor does not receive a Notice of Commencement Date
(g) (i)   the Employer fails to comply with Sub-Clause 1.6 (Contract Agreement)

Items falling under Category B above
(g) (ii)   the Employer assigns the Contract without the required agreement under Sub-Clause 1.7 (Assignment)
(h) a prolonged suspension affects the whole of the Works as described in Sub-Clause 8.12 (Prolonged Suspension)
(i) the Employer becomes bankrupt or insolvent
(j) the Employer is engaged in corrupt, fraudulent, etc. practice

16.2.2 Termination

Before taking action under this sub-clause, it would be appropriate for the Contractor to obtain legal advice to ensure that the giving of a Notice is legally sound.

(a) Unless the Employer remedies any matter described in the Contractor' 14 days of receiving the Notice, the Contractor may terminate the Contract by giving a second Notice to the Employer. The date of termination shall be the date on which the Employer receives this second Notice.

(b) In the case of any matter referred to in Sub-Clause 16.2.1 Category B above, the Contractor, by giving a Notice to the Employer, may immediately terminate the Contract. The date of termination shall be the date on which the Employer receives this Notice.

If the Contractor suffers delay or incurs cost in the 14 days period ((a) above), then the Contractor shall be entitled to payment of Costs with Profit and EOT.

16.3 Contractor's Obligations After Termination

After termination of the Contract under

- Sub-Clause 15.5 (Termination for Employer's Convenience)
- Sub-Clause 16.2 (Termination by Contractor)
  or
- Sub-Clause 18.5 (Optional Termination)

the Contractor shall promptly

(a) cease all further work, except for such work instructed by the Engineer for the protection of life or property or safety of the Works. The Contractor is entitled to payment for this work including profit by reference to Sub-Clause 20.2.
(b) deliver to the Engineer all Contractor's Documents, Plant (apparatus, equipment, machinery, etc. intended to form part of the Permanent Works as defined in Sub-Paragraph 1.1.65) and other work for which the Contractor has received payment
(c) remove all other Goods (defined in Sub-Paragraph 1.1.44) from the Site except as necessary for safety and leave the Site.

The termination of a Contract is a traumatic event for all engaged on Site. However, it is important that the hand-over of the Plant and removal of the Goods should take place in an orderly manner under the supervision of competent staff. Security arrangements shall be maintained until the Contractor has removed his Goods from Site. Thereafter, the responsibility passes to the Employer.

16.4 Payment after Termination by the Contractor

'After Termination under Sub-Clause 16.2 (Termination by Contractor), the Employer shall promptly

(a) *pay the Contractor in accordance with Sub-Clause 18.5 (Optional Termination) and*
(b) *subject to the Contractor's compliance with Sub-Clause 20.2 (Claims for Payment and/or EOT) pay the Contractor the amount of any loss of profit or other losses suffered by the Contractor as a result of this termination'.*

In addition to the above, the routine Measurement and Valuation (Clause 12) including Variations, Claims, etc. have also to be finalised and agreed.

17

Clause 17 Care of the Works and Indemnities

This clause was titled 'Risk and Responsibility' in the First Edition, and in this edition, it has been re-written and re-titled.

The separation of Employer's Risks and Contractor's Risks is clearly stated and a new sub-clause, Sub-Clause 17.6 (Shared Indemnities) has been added.

17.1 Responsibility for Care of Works

This sub-clause provides the basic requirement that the Contractor shall be fully responsible for the care of the Works, Goods and Contractor's Documents from the Commencement Date until the Date of Completion of the Works (i.e. the date stated in the Taking-Over Certificate issued by the Engineer). On receipt of the Taking-Over Certificate, the Contractor will advise the Insurance provider and responsibility for the Works will pass to the Employer.

However, the Contractor remains responsible for the care of any outstanding work until completed. The Contractor should ensure that the satisfactory completion of outstanding work is signed off by the Engineer. The date of the handover is to be agreed and recorded.

17.2 Liability for Care of the Works

The Contractor remains responsible for any loss or damage caused by the Contractor after the issue of the Taking-Over Certificate. This continuing responsibility could arise if the Contractor damaged any part of the Works or the repair of previously damaged work which occurred before of the issue of the Taking-Over Certificate and for which the Contractor was responsible.

The Contractor shall have no liability whatsoever by way of indemnity or otherwise, for loss or damage or damage to the Works, Goods, or Contractor's Documents caused by any

Guide to the FIDIC Conditions of Contract for Construction: The Red Book 2017, First Edition.
Michael D. Robinson.
© 2023 John Wiley & Sons Ltd. Published 2023 by John Wiley & Sons Ltd.

of the following events (excepting to the extent that such Works, Goods, or Contractor's Documents have been rejected under Sub-Clause 7.5 (Defects and Rejection):

(a) interference, whether temporary or permanent, with any right of way, light, air, water, or other easement (other than that resulting from the Contractor's method of construction) which is the unavoidable result of the execution of the Works in accordance with the Contract

(b) use of occupation by the Employer of any part of the Permanent Works except as may be specified in the Contract

(c) fault, error, defect, etc. in any element of the design… *'which an experienced contractor… would not have discovered when examining the Site and the Specification and Drawings…'.* It has been long pointed out that given the limited time, the Contractor is given to prepare his tender offer, the possibility of the Contractor identifying errors in the Employer's design are limited. Additionally, the Contractor's staff are not necessarily experienced designers as are the Employer's designers. Any errors and omissions detected in the preparation of tenders are conventionally addressed either in the official site visits prior to tender or by written correspondence addressed to the Employer.

(d) the Contractor is liable for the consequences of forces of nature which are allocated to him in the Contract Data. It can be expected that the dominant force of nature identified in the Contract Data will relate to hydraulic issues (river flows, cofferdam designs, monsoon weather etc.). The insurance provider will have his own engineering team to advise him in relation to these issues and can be expected to request a significant amount of detailed information from both the Employer and the Contractor.

A further source of difficulty can arise from heavy rainstorms including monsoon weather. It is advisable that the Contractor identifies official weather stations in the general area of a given project and obtains past weather data from those weather stations that can be studied, and the potential risk assessed. Records of past river flows provide essential information to the insurers, designers, and those engaged in the actual construction.

A decision whether an experienced contractor could reasonably be expected to have taken adequate precautions is often largely based on historical records.

(e) any of the events or circumstances identified in Sub-Clause 18.1 (Exceptional Events)

(f) any act or default of the Employer's personnel.

If any of the events listed above occur and cause damage, the Contractor shall promptly give a Notice to the Engineer. Thereafter the Contractor shall repair the damage in accordance with the Engineer's instructions. However, it frequently happens that an individual event may happen very quickly, requiring emergency measures to be taken without delay to protect lives and property.

In such an event, Contract formalities may have to be suspended temporarily. The Contractor should keep all appropriate records and liaise very closely with the Engineer in any on-going emergency.

The Contractor shall, subject to Sub-Clause 20.2 (Claims for Payment and/or EOT), be entitled to payment of cost and award of an EOT. Should the Contractor be partly responsible for the cause of the event, then a proportional payment only shall be made to the Contractor.

Note:
Sub-Clause 18.4 (Exceptional Events) lists 'Exceptional Events' as not foreseeable events at Base Date and are to be administered in a different manner to those events described above.

17.3 Intellectual and Industrial Property Rights

This sub-clause provides protection to each Party in respect of breaches of copyright or other intellectual or industrial property rights.

If a Party receives a claim but fails to give a Notice to the other Party of the claim within 28 days of receiving it, then the first Party has waived his rights.

The Employer shall indemnify the Contractor in respect of:

(a) unavoidable result of the Contractor's compliance with the Specifications, drawings, and Variations
(b) a result of the Works being used by the Employer
  (i) for a purpose indicated or inferred from the Contract
  (ii) in conjunction with anything not supplied by the Contractor (unless previously disclosed before the Base Date)

The Contractor will indemnify the Employer in respect of:

- the Contractor's execution of the Works
- the use of the Contractor's Equipment

The indemnifying Party has overall responsibility for negotiating the settlement of the claim (including the cost of litigation) at his own cost. The other Party is required to provide assistance at the request and cost of the indemnifying Party and shall not make admissions or take actions which might prejudice the rights of the indemnifying Party.

17.4 Indemnities by Contractor

The Contractor indemnifies the Employer from third party claims in respect of:

(a) bodily injury, sickness, disease, or death of any person arising from the Contractor's execution of the Works unless attributable to any wilful act or breach of the Contract by the Employer, the Employer's Personnel, or any of their respective agents
  or
(b) damage to or loss of property (real or personal) (other than the Works) which arises by reason of the Contractor's execution of the Works and is attributable to any negligence, wilful act, or breach by the Contractor.

In addition, the Contractor will indemnify the Employer against all acts, errors, or omissions by the Contractor in carrying out the Contractor's design obligations (if any) that results in the Works (or Section or Part or major item of Plant), when completed, not being fit for their intended purpose.

17.5 Indemnities by Employer

The Employer indemnifies the Contractor and his staff from third party claims in terms of

(a) bodily injury, sickness, disease, or death which is attributable to any negligence, wilful act or breach of the Contract by the Employer
or
(b) damage to or loss of any property (other than the Works) to the extent that such damage or loss arises out of any event described under sub-paragraph (a) to (f) of Sub-Clause 17.2 (Liability for Care of the Works)

17.6 Shared Indemnities

This provision for 'Shared Indemnities' is introduced for the first time in this Second Edition.

The Contractor's liability to indemnify the Employer under Sub-Clause 17.4 (Indemnities by Contractor) and/or under Sub-Clause 17.3 (Intellectual and Industrial Property Rights) shall be reduced proportionately to the extent that any event described under sub-paragraphs (a) to (f) of Sub-Clause 17.2 may have contributed to the said damage, loss, or injury.

Similarly, the Employer's liability to indemnify the Contractor under Sub-Clause 17.5 (Indemnities by Employer) shall be reduced proportionately to the extent described under Sub-Clause 17.1 (Responsibility for Care of the Works) and/or under Sub-Clause 17.3 may have contributed to the said damage, loss, or injury.

Claims falling under the above headings are uncommon. The Employer and Contractor have a joint interest in disputing any third-party claim and the amount thereof. Should amounts have to be paid to a third party and the parties cannot agree proportionate responsibility, it may be mutually convenient for the DAAB to give their opinion in the matter as described in Sub-Clause 21.3 (Avoidance of Disputes).

18

Clause 18 Exceptional Events

In the First Edition, this clause was titled 'Force Majeure' and numbered Clause 19. However, the general content of this clause is similar to that previously provided.

18.1 Exceptional Events

An 'Exceptional Event' means an event or circumstance which
'(i)   *is beyond a Party's control*
(ii)   *the Party could not reasonably provide against before entering into the Contract*
(iii) *having arisen, such Party could not have reasonable arisen or overcome and*
(iv) *is not substantially attributable to the other Party'.*

Provided that the above conditions are met, an Exceptional Event may comprise but not be limited to:

(a) *war, hostilities, invasion act of foreign enemies*
(b) *rebellion, terrorism, revolution, insurrection, military or usurped power or civil war*
(c) *riot, commotion by persons other than the Contractor's personnel or other employees of the Contractor or subcontractors*
(d) *strike or lock-out not solely involving the Contractor's personnel and other employees of the Contractor and subcontractors*
(e) *encountering munitions of war, explosive materials, etc. excepting as may be attributable to the Contractor's activities*
(f) *natural catastrophes such as earthquakes, hurricanes, etc.*

The relationship of this sub-clause to Sub-Clause 17.2 (Liability for Care of the Works) is to be noted.

Guide to the FIDIC Conditions of Contract for Construction: The Red Book 2017, First Edition.
Michael D. Robinson.
© 2023 John Wiley & Sons Ltd. Published 2023 by John Wiley & Sons Ltd.

18.2   Notice of an Exceptional Event

If a Party is prevented from performing any obligations under the Contract due to an Exceptional Event, then the affected Party (which is likely to be the Contractor) shall give a Notice to the other Party, detailing the Exceptional Event. The Notice shall describe the activities which are hindered or prevented because of the occurrence of an Exceptional Event. The Notice is to be given within 14 days after the affected Party became aware of the Exceptional Event and the affected Party is excused further performance until the affected Party can safely recommence normal operations.

Should the affected Party give the Notice after the stated period of 14 days, then the affected Party shall only be excused further performance from the date of the delayed Notice. Notwithstanding all of the above, the affected Party shall continue to perform those operations not affected by the Exceptional Event.

18.3   Duty to Minimise Delay

Each Party shall at all times use all reasonable endeavours to minimise any delay arising as a result of an Exceptional Event.

Exceptional events may arise with little prior warning and the situation is likely to change quickly. Having given the Notice described above, the Contractor may provide the other Party with periodic reporting of the actual situation and of the Contractor's activities to minimise the consequences of the Exceptional Event. This Sub-Clause indicates that the affected Party shall *'describe the effect (or any delay) each 28 days'.* This would appear to be too long an interval for accurate record keeping which will be required for evaluation of both cost and EOT as described in Sub-clause 18.4 below.

18.4   Consequences of an Exceptional Event

This sub-clause affirms that if the Contractor (assumed to be the affected Party) suffers delay and/or incurs cost as a consequence of the Exceptional Event of which he gave Notice under Sub-Clause 18.2 (Notice of an Exceptional Event), he is entitled to EOT, and payment of Cost provided that the Exceptional Event occurred in the country. On occasions an Exceptional Event which occurs in another country may cause delays (e.g. supply of fuels or material may be disrupted). This is an inconvenience to the Contractor, however, no remedy is provided in these FIDIC forms of Contract.

The evaluation of the Contractor's costs will be a difficult task for the Contractor to prepare and for the Engineer to check, even if accurate records are made of the standing work hours of both manpower and goods normally employed in the Works but made idle by the Exceptional Event. In particular, the evaluation of cost/hour will be difficult to assess, requiring a detailed review of any cost data in the Contractor's possession. It will be noted that this sub-clause does not provide for the payment of profit to the Contractor.

18.5 Optional Termination

If the execution of a substantial part of all the Works is prevented for a continuous period of 84 days by reason of an Exceptional Event or for multiple periods which total more than 140 days due to the same Exceptional Event, then either Party may give a Notice of Termination of the Contract. This situation would be extreme and is unlikely to take place without consultations between the Parties, as the ramifications of such termination may have serious consequences elsewhere.

Should a Notice of Termination be issued, the date of the termination shall be seven days after receipt of the Notice by the other Party. Thereafter, the Contractor shall as soon as possible submit detailed supporting particulars of the value of work done, as follows:

(a) the amounts payable for any work for which a price is stated in the Contract
(b) the cost of Plant and Materials ordered for the Works which have been delivered or which the Contractor is obliged to take delivery. The Plant and Materials shall become the property of the Employer following payment for the same to the Contractor
(c) any other Cost or liability which in the circumstances was reasonably incurred in expectation of completing the Works. This may be referred to as a 'catch all' clause. Many items required for temporary works may be on site, e.g. aggregates, cement, etc. The Contractor's stores are likely to contain a significant amount of project specific materials and spare parts requiring disposal at a loss.
(d) the cost of removal of Temporary Works and Contractor's Equipment from the Site and the return of these items to the Contractor's home country (or countries if the Contractor is a JV). The Contractor may decide to send all or part of the Temporary Works and Contractor's Equipment to another project in another country, but the cost claimed under this heading shall not exceed the cost of return to the Contractor's home country.
(e) the cost of repatriation of Contractor's staff and labour employed *wholly* in connection with the Works at the date of termination.

The Engineer shall then proceed under Sub-Clause 3.7 (Agreement or Determination) to agree or determine the value of work done. It is further stated that *'the date the Engineer receives the Contractor's particulars under this sub-clause shall be the date of commencement of the time limit for agreement under Sub-Paragraph 3.7.3'*. Sub-Paragraph 3.7.3 states in part *'The Engineer shall give the Notice of agreement, if agreement is achieved, within 42 days or within such other time limit as may be proposed by the Engineer and agreed by both Parties. . . '*.

However, the time required for evaluation and assessment of Contractor's Cost of termination will depend on two principal factors:

1. The size and complexity of the project
   Smaller, less complex projects are likely to be easier and more quickly evaluated. In comparison, a large, more complex project, which has been under construction for a lengthy time period will be more difficult to evaluate and more time will be required to make an evaluation. The assistance of specialist cost accountants may also be required.

2. The progress status of the Works
   Contracts which have been in progress for a short time will be less developed and there-fore easier to shut down. Consequently, the evaluation of Cost of termination will be easier to achieve. In comparison, a large project, which has been in progress for a lengthy period of time, will be more difficult to close down, not least for security rea-sons. The assessment of Contractor's costs and subsequent agreement of those Costs are likely to take far more than 42 days stated in Sub-Paragraph 3.7.3. This issue will require discussions between Engineer and Contractor. The fixing of a precise timetable for eval-uating costs may have to be delayed for an indeterminate amount of time until the head-ings and scope of the evaluation is clearer.

Sub-Clause 18.5 further states 'the Engineer shall issue a Payment Certificate for the amount agreed or determined, without the need for the Contractor to submit a Statement'. Considering that the assessment of Contractor's Costs may take many months to complete and agree, it would be appropriate if the Engineer were to issue more than one Payment Certificate corresponding to the amounts agreed in (say) monthly intervals.

18.6 Release from Performance Under the Law

'In addition to any other provision of this Clause if an event arises outside the control of the Parties, including but not limited to an Exceptional Event which

(a) *makes it impossible or unlawful for either Party (or both Parties) to fulfil their contractual obligations, or*
(b) *under the law governing the Contractor entitles the Parties to be released from further performance of the Contract'*

and if the Parties are unable to agree an amendment to the Contract, then after one Party gives a Notice to the other Party, the Contract may be terminated. The Contractor shall be paid as described in Sub-Clause 18.5.

Events which could trigger application of this Sub-Clause could include:

- an abrupt change in the political direction in the host country
- significant worldwide economic and fiscal downturn
- a pandemic or an outbreak of a major health issue

Should the Contract be terminated in exceptional circumstances as noted above, the Contractor will be concerned that he may not receive payments rightfully due to him.

Note:
Several countries are able to offer contractors based in those countries' special export insur-ance against the possibility of a foreign State Employer failing or being unable to pay con-tractors the amounts due. Notably this situation arose in the aftermath of the Iran-Iraq war of the 1980s, when the Iraqi government was unable to pay the amounts due to foreign contractors. After due legal process, many contractors were able to access the state spon-sored insurance available in their own countries.

19

Clause 19 Insurance

Note:
Sub-Clause 19.1, which follows, requires that insurances will be provided and managed by the Contractor. In the Contract Data, there are nine items referring to insurances which have to form part of the insurance package to be provided by the Contractor.

There is no reference to any insurance to be provided by the Employer. However, occasionally the Employer may choose to provide insurances. The Employer is most likely to provide insurances when the contract is one of the number of contracts required for the construction of a very large project. The Employer may consider that he can obtain such insurances at more favourable rates and that the insurances can be tailored to best meet all of his requirements. In the absence of information to the contrary, the Contractor would be entitled to assume that the Employer-provided insurances would be compliant with this Sub-Clause 19.1. Of immediate concern to a tenderer would be the amount of excess to be deducted from any insurance payments.

19.1 General Requirements

The opening sentence of this Sub-Clause states that '. . . *the Contractor shall effect and maintain all insurances for which the Contractor is responsible with insurers and in terms, both of which are subject to consent by the Employer'*. (Underscoring added)

The time interval between the issue of the Tender Documents and submittal of tenders is not fixed in these Conditions of Contract. However, the Contractor must have completed all arrangements with his chosen insurer and have agreed the cost of the insurance policy for inclusion in his tender offer. In addition, the above quoted underscored extract from Sub-Clause 19.1 requires the Contractor to have obtained the consent of the Employer to both the insurer and the terms of any policy also before a Letter of Acceptance is issued. The Contractor, having selected his proposed insurer, will provide a draft insurance proposal in conformity with the Conditions of Contract with terms indicated together with costing for inclusion in his tender offer. Should the tender prove successful, the finer detail of the insurance may be readily agreed. The FIDIC Conditions of Contract were originally intended for use on projects in other countries where the local conditions of contract

Guide to the FIDIC Conditions of Contract for Construction: The Red Book 2017, First Edition.
Michael D. Robinson.
© 2023 John Wiley & Sons Ltd. Published 2023 by John Wiley & Sons Ltd.

(if any) were unsuitable for major projects. Financing in foreign currencies is increasingly provided by international agencies (World Bank, etc.) enabling foreign contractors to be paid in convertible currencies. Consequently, the type of insurances indicated in this edition will likely be provided by a reputable offshore provider whose premiums and settlements will also be valued in a convertible currency.

However, in more recent years the FIDIC Conditions of Contract have been increasingly specified for use on infrastructure projects, notably in Eastern Europe. These have been used largely financed by the EU and valued in Euros. Local contractors have successfully bid for these contracts. However, many of these contractors have difficulty in understanding and providing project insurances conforming to the FIDIC model. In addition, State Employers also face difficulty in assessing these insurance proposals. The outcome is that where a contractor has failed, the insurance policy has frequently proved inadequate and in a number of instances of no value.

It is suggested that an alternative version of Clause 19 be developed to better suit the conditions highlighted above.

Proof of payment of Insurance Premiums

This sub-paragraph states, *'the Contractor shall produce the insurance policies . . .'* and *'the Contractor shall . . . submit a copy of each receipt of payment to the Employer (with a copy to the Engineer)'*. This requirement can be met by providing the Employer with an authenticated copy of the insurance policy in its final approved form. Copies of receipts of payment provided by the insurers are best directed through the Contractor to the Employer.

Failure to maintain policies

Should the Contractor fail to pay insurance premiums, the Employer may himself pay the premiums and back-charge the Contractor (possibly by deduction from interim payments). The danger of this procedure is that the failure of the Contractor to pay premiums may be an indicator of financial problems. A safeguard would be to require the insurer to immediately notify the Parties of any failure on the part of the Contractor.

Compliance

'If either the Contractor or the Employer fails to comply with any condition of the insurances, the Party so failing shall indemnify the other Party against all direct losses and claims . . .'.

Notification of Changes

The Contractor is responsible for notifying the insurers of any changes in the execution of the Works and the adequacy and validity of the insurances through the performance of the Works.

These are routine matters to be attended by the Contractor:

(1) The insurer may require a periodic report from the Contractor detailing not only the physical progress of the Works but its financial status, including details and values of varied works, additional works which are likely to change the final valuation of the Works. The arrival and departure dates of the Contractor's major items of Equipment to and from the Site should be included in the report, as premiums may vary according to the actual value of the Contractor's Equipment on Site.

(2) On larger projects the insurer may send a delegate to the Site (at say 3 months intervals) to review the status of the Works at first hand. During this visit, the delegate will collect from both Parties any data required for settlement of existing claims. The Employer may also wish to meet with the delegate.

Shared liability

Where there is a shared liability, the loss shall be borne by each Party in proportion to the liability of each Party. Should the non-recovery from the insurers arise because of a breach of this clause by one Party, then the defaulting party shall bear the loss suffered.

19.2   Insurance to be provided by the Contractor

The Contractor shall provide the following insurances:

19.2.1   The Works

(Works defined as 'the Permanent Works and Temporary Works')
   The Contractor is required to insure in the joint names of the Contractor and Employer the following:

(a) the Works and Contractor's Documents, together with Materials and Plant for incorporation in the Works for their full replacement value. This insurance cover shall extend to include loss or damage of any part of the Works which arise as a consequence of a failure of elements designed or constructed with defective material or workmanship.
(b) an additional amount of 15% or such other amount as may be specified in the Contract Data. This amount is intended to cover any additional costs incidental to the rectification of loss or damage.

   This insurance shall cover against loss or damage from whatever cause until the issue of the Taking Over Certificate. The insurance shall continue to the date of issue of the Performance Certificate in respect of any damage caused by the Contractor in executing his remaining obligations after the issue of the TOC. The insurance cover specifically excludes wear and tear, shortages, and pilferages after Taking Over by the Employer.
   'The insurance cover to be provided by the Contractor may exclude any of the following:

(1) *the cost of making good any part of the Works which is defective'*
   Contractors routinely correct their own defective work, particularly if the costs of rectification are minor. However, there may occur a major incident where rectification costs may be high. Consequently, an insurance policy is likely to have a 'minimum value claim' clause, whereby the Contractor may only claim for claims above an agreed minimum value.
'(2) *indirect or consequential loss or damage, including any reductions in Contract Price for delay'.*
   It is likely that a similar insurance arrangement as described in (1) above suffice. Insurances in respect of loss of revenue for delay are likely to be prohibitively expensive.

'(3) *wear and tear, shortages and pilferages*'
    A similar approach as (*1*) is likely.
'(4) . . .*risks arising from Exceptional Events*'
    It would be necessary for an assessment to be made of the likelihood of events requir-
    ing insurance cover for any possible 'Exceptional Events'.

19.2.2 Goods (Contractor's Equipment, Materials, Plant, and Temporary Works)

The Contractor shall insure (in joint names of the Contractor and the Employer) the
(Contractor's) Goods for their full replacement value including delivery to Site. The wording 'full replacement value' means that the insurance term 'like for like' will apply in any
settlement of claim.

This sub-clause provides the possibility that the Employer may specify the amount of this
insurance in the Contract Data. However, the Contractor will have a full evaluation of the
amounts paid by him for the various elements of Goods. Should an insurance claim be
made in respect of damage to an element of Goods, the insurer will negotiate with the
Contractor and offer the replacement value in any settlement. As the Works (or Sections)
approach conclusion, the Contractor will request to progressively remove from Site the
Goods which are no longer required on Site for use for refurbishment and reallocation
elsewhere. The insurer is to be kept informed of these disposals and will modify the insurance package (and its cost) accordingly.

19.2.3 Liability for Breach of Professional Duty

This sub-paragraph is applicable only if the Contractor is responsible for the design of a
part of the Permanent Works. A review of the Tender Documents will reveal if the
Contractor has any such design responsibility. The amount of any insurance is to be provided in the Contract Data, Page 6. The insurance shall indemnify the Contractor against
liability should the Contractor's design not be fit for the intended purpose.

19.2.4 Injury to Persons and Damage to Property

The Contractor is required to insure (in the joint names of the Contractor and the Employer)
against liabilities for death and injury to any person arising out of the performance of the
Contract.

19.2.5 Injury to employees

The Contractor shall insure his employees in respect of injury, sickness, disease, or death
of any person employed by the Contractor and any of the Contractor's Personnel (which
includes the personnel of Subcontractors). The Subcontractor may choose to provide a
separate insurance for his personnel, but the Contractor is responsible for his compliance.

The Contractor shall have a clear policy on these issues to be followed by all subcontractors. It is also possible that there may already exist State sponsored insurance/medical insurance scheme available to all local employees engaged on the Site. In addition to paying the subscriptions to these services, the Contractor may consider providing additional insurance benefits as this usually will be appreciated by the local employees.

Expatriate personnel will have accident/medical insurance provided as part of their employment package, including a medivac provision. Depending on the scale of the project and numbers of personnel on site, a site hospital may be provided in addition to the standard first aid facilities.

19.2.6   Other Insurances Required by Law and by Local Practice

This is a reminder to the Contractor that the local law (or custom) may require the Contractor to arrange other insurances. Such insurances will likely be separate to the principal insurances described in this Clause 19. Premiums will require to be paid in local currency. Any such obligations are to be researched by the Contractor's tender team and the cost included in the Tender price.

Other Considerations:
(1) It is likely that the Parties will open a joint bank account to receive payments from the insurers. Distributions from this account will require authorisation by both Parties. It is intended that monies received from the insurers will be used to rectify the losses or damages to which the payment refers.
(2) Neither Party should damage the interests and rights of the other Party in respect of any claim notified to the Insurers. It may happen that the interests of the Employer and Contractor differ. Each Party should proceed with caution and avoid attributing blame until such time as the insurers have formally responded to the insurance claim. The Contractor is obliged to promptly notify the insurer of *'an event which may give rise to a claim under the insurance policy'*. The Contractor is required to advise the Insurer of *'the facts of the* matter' and not express opinions as to cause. It is the obligation of the Insurer to determine the cause of the event giving rise to the claim. The Employer and the Engineer are also required to limit themselves to giving *'matters of* fact' only. The insurers will decide if a claim is valid under the terms of the insurance policy and advise the Contractor accordingly. Detailed negotiations concerning the value of the valid claim can then commence.

20

Clause 20 Employer's and Contractor's Claims

Notes:

(a) The Second Edition contains significant changes from the arrangements included in the First Edition. Notably, in the First Edition there was limited specific reference to the entitlement of the Employer to make claims against the Contractor, other than for Delay Damages.

In this Second Edition, the Employer may make claims against the Contractor in much the same manner as the Contractor makes claims against the Employer.

Although the authority of the Engineer is confirmed and strengthened in the Second Edition (reference Clause 3 generally and Sub-Clause 3.7 specifically), it remains that the Engineer is an employee of the Employer and is assigned to fairly evaluate the respective claims of both the Employer and the Contractor. Therefore, the Engineer cannot participate in the preparation and submittal of either the Employer's or the Contractor's claims. The criteria adopted for the evaluation of claims must be consistent.

(b) Reference is made to the FIDIC Document 'The FIDIC Golden Principles. First Edition 2019'. (This is available online free of cost and a reading of this document is recommended.)

Golden Principle GP1 on Page 8 of this FIDIC document states:

'*GP1: The duties, rights, obligations, roles and responsibilities of all the Contract Participants must be generally as implied in the General Conditions, and appropriate to the requirements of the project*'.

and

'*the following are examples of modifications that do not comply with GP1:*

– *Under a Red Book, or Yellow Book contract, the Engineer is required to obtain the Employer's approval before making any determination of a Contractor's claim or granting any extension of time pursuant to Sub-Clause 3.7 (or Sub-Clause 3.5 in the 1999 Editions). The Engineer's role as defined in a FIDIC Contract is to fairly determine the Contractor's entitlements in accordance with the Contract conditions, and this should not be subject to influence or control by the Employer. If the Employer disagrees with the Engineer's determination, the Contract provides an avenue for resolving this by the Dispute Avoidance/Adjudication Board*'.

Guide to the FIDIC Conditions of Contract for Construction: The Red Book 2017, First Edition.
Michael D. Robinson.
© 2023 John Wiley & Sons Ltd. Published 2023 by John Wiley & Sons Ltd.

The effects of restrictions on the power and authority of the Engineer to take actions leading to additional payment to the Contractor are reported elsewhere. This issue is unlikely to be resolved soon because in a number of jurisdictions it is unlawful for additional payment to be made in the manner given in these Conditions of Contract. However, it is not correct that some Employers employ these restrictive practices when it is not necessary and not required by the local law.

However unsatisfactory the current situation may be, the Engineer and the Contractor can only proceed with the administration of a Contract as it is written. The Contractor may elect to add potential additional financial charges in his tender, which may put him at a disadvantage to his competitors. Alternatively, it would be beneficial if the Engineer and Contractor were to informally cooperate to identify at an early stage those site issues that will likely result in additional payments becoming due to the Contractor. This knowledge may assist the Employer organise his financial planning in a more cohesive manner.

20.1   Claims

The following categories of claim are provided:

(a) Employer Claims against the Contractor for any additional payment (or reduction in price) or to an extension of the DNP
(b) Contractor Claims against the Employer for additional payment and/or EOT

Sub-Clause 20.2 (Claims for Payment and EOT) shall apply to these categories of claim. A further category of claim has now been introduced:

(c) Should either Party consider that he is entitled to another entitlement or relief against the other Party, (which is not the subject of (a) or (b) above), then he may give a Notice to the Engineer, copied to the other Party, detailing the requested entitlement or relief. Such other entitlement may be of any kind whatsoever.

No examples are provided of this claim category which are assumed to be referring to what might be termed 'sundry item not specifically related to the performance of the contract' and which are not specifically addressed in the Contract Documents and yet do arise from time to time during the execution of the Works.

If the claim under reference cannot be resolved or agreed between the Parties, or if the Engineer disagrees with the claimant Party's request (or do not respond), then the claimant Party may give Notice referring the claim to the Engineer and Sub-Clause 3.7 (Agreement or Determination) shall apply. There is no time limit for submitting this Notice, but it should be given as soon as practical.

20.2   Claims for Payment and EOT

The general opening sentence of this sub-clause specifically refers to Sub-Clause 20.1 above and states that the following claim procedure shall apply.

20.2.1 Notice to Claim

Compliance with the requirements of the following sentence is fundamental to any claim to be made by the Contractor or the Employer. Underscoring has been added for emphasis.

'*The claiming Party shall give a Notice to the Engineer describing the event or circumstance giving rise to the cost, loss or delay or extension of the DNP, for which the Claim is made as soon as practical, and no later than 28 days after the claiming Party became aware, or should have become aware, of the event or circumstance*'.

Should the claiming Party fail to comply with the above requirement, then the potential claim is rendered nil and void. There is a permissive clause contained in Sub-Clause2.2. that would allow the Engineer to extend the period of validity for the claim notification, However, it would be prudent if the claimant were not to rely on this extension being granted.

In event that the potential claimant Party is unsure if an event or circumstance has developed to a point where a formal claim is justified, it would be appropriate to give a formal Notice of Claim to avoid the potential claim being made out of time. A claim can always be withdrawn should circumstances change and a formal claim is no longer justified.

20.2.2 Engineer's Initial Response

'*If the Engineer considers that the claiming Party has failed to give a Notice of Claim within the period of 28 days, then the Engineer shall, within a further period of 14 days, give a Notice to the claiming Party with reasons*'.

It is presumed that this Notice would confirm that the claim is out of time and will not be considered further.

Should the Engineer not give such a Notice within the 14-day period, the Notice of Claim shall be deemed to be a valid notice. This indicates that the Engineer does not have the authority to reject the claim in exceptional circumstances. However, if the other Party disagrees with the decision of the Engineer, he is entitled to give a Notice to the Engineer requesting a review of the reasons why such late submission is not justified.

20.2.3 Contemporary Records

This is an additional definition (not given in the First Edition) and means '*records that are prepared or generated at the same time, or immediately after the event or circumstance giving rise to the claim*'. This definition would include daily site records, laboratory data, survey data, weather records, etc. which would be maintained by the Contractor. The Employer may decide to monitor these daily records and may instruct the Contractor to keep additional records. Copies of these records are to be provided to the Engineer if requested. (It is likely that these records will be provided to the Engineer as part of the claim presentation and are likely to be required in the event of any referral to the DAAB.)

20.2.4 Fully detailed claim

A fully detailed claim is defined as a submission which includes

(a) '*a detailed description of the event or circumstance giving rise to the Claim*
(b) *a statement of the contractual and/or legal basis of the Claim*'

This statement will be derived from any document forming part of the Contract Documents, the Engineer's instructions, contemporary records, external documents including central and local government regulations, external hydrological records, etc.

(c) all contemporary records (reference Sub-Paragraph 20.2.3 (Contemporary Records))

(d) (i) detailed supporting particulars of the additional payment and EOT claimed by the Contractor or

(ii) detailed supporting particulars of the amount of reduction of the Contract Price or extension of the DNP, claimed by the Employer.

Within either

'(i) *84 days after the claiming Party becomes aware, or should have become aware, of the event or circumstance giving rise to the Claim,*
or

(ii) *such other period (if any) as may be proposed by the claiming Party and agreed by the Engineer*

The claiming Party shall submit to the Engineer a fully detailed claim'.

Should the claiming Party fail to submit the Statement under sub-paragraph (b) within the time limit of 84 days, the Notice of Claim shall be deemed to have lapsed and the Engineer shall, within a further period of 14 days give a Notice to the claiming Party accordingly.

If the Engineer does not give such a Notice within a period of 14 days, the original Notice of Claim shall be deemed to be a valid Notice. (By what criteria the Engineer shall or shall not issue a Notice of an elapsed claim is not stated.)

If the other Party disagrees with the deemed valid Notice of Claim, he shall in turn issue a Notice to the Engineer giving details of the reasons why he disagrees with the Engineer. Thereafter the agreement or determination of the Claim under Sub-Paragraph 20.2.5 shall include a review by the Engineer of such disagreement.

20.2.5 Agreement or determination of the Claim

Sub-Clause 3.7, which is headed 'Agreement or Determination' describes how the Engineer shall proceed to *'agree or determine'* a valid claim presented by either Party.

The contents of this Sub-Paragraph 20.2.5 complements the requirements of Sub-Clause 3.7, notably on the matter of Agreement or Determination by the Engineer.

The user of these Conditions of Contract is advised to take care that the giving of a Notice and the various time limitations of both sub-clauses are strictly observed.

20.2.6 Claims of Continuing Effect

Claims of continuing effect can be characterised by events and circumstances such as:

(a) Increased labour costs as a result of statutory increases in labour wages and benefits

(b) Increased cost of locally obtained materials including fuels, cement, bitumen, etc., where prices are set by the Government

(c) Additional or increased import taxes (not applicable if the project is described as 'free of local taxes'.

These additional costs can be readily evaluated on a monthly basis from labour records and purchasing office/stores data.

This sub-clause states that:

(a) the fully detailed claim submitted under Sub-Paragraph 20.2.4 shall be considered as interim

(b) the Engineer shall respond to the contractual or legal basis of the Claim by giving a Notice to the claiming Party (generally 42 days if not agreed otherwise)

(c) the valuation of the Claim shall be updated at monthly intervals and submitted to the Engineer for approval and inclusion in the next IPC

(d) a final Claim valuation within 28 days of the end of the effects of the event or circumstance.

It is possible that events or circumstances may arise which will reduce costs and entitle the Employer to formally present a claim for a reduction in Cost. The necessary data will be available only with the Contractor who will likely be requested to provide the same to the Engineer for his review. The appropriate adjustments can be made in the next IPC.

20.2.7 General Requirements

After receiving Notice of Claim, and until the Claim is agreed or determined under Sub-Paragraph 20.2.5 (Agreement or Determination of the Claim), the Engineer shall include in the next IPC adjustments such amounts for any Claim as have been reasonably substantiated as due to the claiming Party.

It is also stated that *'the Employer shall only be entitled to claim any payment from the Contract or extend the DNP or set off against or make any deduction from any amount due to the Contractor by complying with this Sub-Clause 20.2'*.

The requirements of this Sub-Clause 20.2 are in addition to those of any other sub-clause which may apply to the Claim.

21

Clause 21 Disputes and Arbitration

21.1 Constitution of the DAAB

The arrangements for the appointment and operation of a DAAB (formerly referred to as DAB) are very similar to those included in the First Edition excepting that this Second Edition provides an Annex containing two supplementary documents titled: 'General Conditions of Dispute Avoidance/Adjudication' and 'DAAB Procedural Rules'.

- The DAAB shall be comprised of either three members or a single member (the 'sole member') as stated in the Contract Data. If the number of members is not stated in the Contract Data, then the DAAB will consist of three members unless agreed otherwise by the Parties.
- Each Party shall select one member named in a listing in the Contract Data. The Parties and the two nominated members shall select the third member who shall act as chairperson.
- The DAAB shall be deemed to be constituted on the date that the members (or sole member) have all signed a DAAB Agreement (refer to Appendix titled 'General Conditions of Dispute/Adjudication Agreement', Pages 107–116, which will form the basis of the Employment Contract of a DAAB member).
- The terms of the remuneration of the DAAB board members of member shall be mutually agreed by the Parties when agreeing the terms of a DAAB Agreement. Each Party shall be responsible for paying one-half of any remuneration to the DAAB board member(s). This remuneration shall include the remuneration of any expert whom the DAAB consults. It is assumed that the Parties will give their authority in advance for this additional cost.

  The Parties are to agree which Party will make the full payment of remuneration and any other approved costs to the DAAB member(s). Should the Employer pay the full amount due to the DAAB members, then the Contractor shall include his 50% share of this payment as a credit to the Employer in the next IPC.

Guide to the FIDIC Conditions of Contract for Construction: The Red Book 2017, First Edition.
Michael D. Robinson.
© 2023 John Wiley & Sons Ltd. Published 2023 by John Wiley & Sons Ltd.

21.2 Failure to Appoint DAAB Members

There are a number of scenarios by which one Party may fail to select other DAAB members, thereby preventing the formation of a DAAB. Considering that a DAAB should be appointed and operational as soon as practical after Commencement Date, it is important that any failure to nominate should be addressed without further delay. To this end *'the appointing entity or official named in the Contract Data (page 7) shall at the request of either or both Parties . . . appoint the members of the DAAB'*.

The Contract Data states, 'Unless otherwise stated, the appointing entity shall be the President of FIDIC or a person appointed by the President'.

Each Party shall be responsible for paying 50% of the remuneration of the appointing entity. Arrangement for settlement of this account is likely to follow the same process used to that applied in the payment of the remuneration due to the DAAB members as described above in Sub-Clause 20.1.

21.3 Avoidance of Disputes

'If the Parties so agree, they may jointly request in writing. . . the DAAB to provide assistance and informally discuss and attempt to resolve any issue or disagreement that may have arisen between them'.
and
'If the DAAB becomes aware of an issue or disagreement, it may invite the Parties to make. . . a joint request'.

During their Site visits it may be expected that the DAAB will separately ask each Party if there are any other issues (which may have technical content) which are not formal claims but about which they should be informally briefed.

This intervention provides an ideal vehicle for the Parties to express themselves freely and privately and without a written record. No reaction (formal or otherwise) is due or should be expected from the members of the DAAB. For this process to have maximum value, the Parties should adequately prepare themselves in advance, so that any discussion can be positive and meaningful.

Note:
It is likely that the Chairman of the DAAB will request that the board is (at monthly intervals) provided with information concerning the status of the Works. This could include progress reports (with photographs), any relevant technical developments including change of design, construction problems, key correspondence (but not claim presentations), claim listings with valuations, etc.

The Employer (Engineer) may undertake this task, but the Contractor may also wish to contribute.

21.4 Obtaining DAAB's Decision

Should a dispute remain unresolved having followed the procedures given in the Contract Documents, then either Party may refer the dispute to the DAAB for its decision. The

DAAB is to be kept informed of the status of any unresolved claims made by either of the Parties. However, the DAAB will not be able to efficiently process all referrals unless these are submitted in an orderly, pre-arranged manner which should take into consideration that the DAAB has a time limit of 84 days after the date of referral to give its decision (N.B. a 'day' is a calendar day. Refer Sub-Paragraph 1.1.27).

The preparation of a referral for a DAAB decision may require a significant time to prepare. Consequently, the arrangements for making a referral must be carefully planned.

Implementation of the matters examined above in Sub-Clause 21.3 (Avoidance of Disputes) may be a valuable aid to settlement.

21.4.1   Reference of a Dispute to the DAAB

As a condition precedent to a referral to the DAAB, the process described in Sub-Clause 3.7 (Agreement or Determination) is to be followed. Summarised, this process requires:

3.7.1   The Engineer is to consult with the Parties jointly and separately in an endeavour to reach agreement.

3.7.2   The Engineer shall make a fair determination of the matter or claim and a Notice to both parties of his determination.

3.7.3   The Engineer shall give Notice of agreement if agreement is reached (time limits apply),

3.7.4   The Engineer may correct errors, if any, in his determination.

3.7.5   If either Party is dissatisfied with a determination of the Engineer, he shall give a Notice of Dissatisfaction (NOD) to the other Party, copied to the Engineer. Thereafter either Party may proceed under Sub-Clause 21.4 (Obtaining DAAB's Decision).

The reference of a Dispute to the DAAB requires:

- compliance with Sub-Clause 3.7 (see above) and be made within 42 days of giving (or receiving) a NOD under Sub-Paragraph 3.7.5
- make reference to this sub-clause
- set out the Referring Party's case relating to the Dispute
- be in writing and copied to the Other Party and to the Engineer
- for a three- man DAAB it is deemed to be received by the Chairman of the DAAB

21.4.2   The Parties' Obligations after the Reference

Both Parties shall make available to the DAAB all that they require for the purpose of making their decision on the Dispute.

21.4.3   The DAAB Decision

'The DAAB is required to complete and give its decision within:

(a) 84 days after receiving the reference or

(b) such other period as may be proposed by the DAAB and agreed by both Parties'

The DAAB shall not be obliged to give its decision if there are unpaid invoices outstanding in their favour.

The decision shall be binding on both Parties who shall promptly comply with the decision whether or not a Party gives a NOD.

Whilst the DAAB's decision will concentrate on the contractual merits of the referral and, if appropriate, provide guidelines for evaluation of the claim, as they are unlikely to agree to carry out a forensic examination of voluminous accountancy documents. In this situation, there are two possibilities: a separate group of accountants perform this task on behalf of the Parties with attendant costs or the Parties perform this task themselves.

If the decision of the DAAB requires a payment of an amount by one Party to the other Party, then this amount shall be immediately due and payable.

'The DAAB proceeding shall not be deemed to be an arbitration and the DAAB shall not act as arbitrator(s)'.

21.4.4 Dissatisfaction with DAAB's Decision (NOD)

If either Party is dissatisfied with the DAAB's decision, then:

(a) one Party may give a NOD to the other Party with copy to the DAAB and the Engineer
(b) this NOD shall state it is a Notice of Dissatisfaction with the DAAB's decision and set out the matter of dispute and reason(s) for dissatisfaction
(c) this NOD to be given within 28 days after receiving the DAAB decision.

Should the DAAB fail to give its decision within the 84-day period stated in Sub-Paragraph 21.4.3 (or any extended period if agreed by the Parties), then within 28 days either Party may give a NOD to the other Party.

'Neither Party shall be entitled to commence arbitration of a dispute unless a NOD in respect of that dispute has been given in accordance with this sub-clause'.

Note:
Other Clauses which permit a Party to commence arbitration proceedings are:

- Sub-Paragraph 3.7.5 (last section) – Dissatisfaction with Engineer's Determination
- Sub-Clause 21.7 (first paragraph) – Failure to Comply with DAAB's Decision
- Sub-Clause 21.8 (sub-heading (b)) – No DAAB in place

'If the DAAB has given its decision as to the matter in dispute and no NOD has been given by either Party within 28 days of receiving the DAAB decision, then the decision shall be final and binding.

If the dissatisfied Party is dissatisfied with only parts of the decision, those parts shall be identified in the NOD and severable from the remainder of the decision. The remainder of the decision shall be binding and final'.

21.5   Amicable Settlement

Where a NOD has been given under Sub-Clause 21.4 (Obtaining the DAAB's Decision), both Parties shall attempt to settle the Dispute amicably before commencement of arbitration.

Unless the participating senior staff fully understand the detail of unresolved claims and have full authority to make key decisions, amicable settlement will be difficult. Too often one Party has a fixed position (often established by others elsewhere) which is unalterable and amicable settlement is not achievable. Consequently, the requirement for seeking an amicable settlement is reduced to a 'box-ticking' exercise as a prelude to arbitration proceedings.

21.6   Arbitration

Unless settled amicably and subject to compliance with:

- Sub-Paragraph 3.7.5 (Dissatisfaction with the Engineer's Determination)
- Sub-Paragraph 21.4.4 (Dissatisfaction with DAAB's Decision)
- Sub-Clause 21.7 (Failure to Comply with DAAB's Decision)
- Sub-Clause 21.8 (No DAB in Place),

any decision of the DAAB which has not become binding shall be finally settled by international arbitration under the Rules of Arbitration of the International Chambers of Commerce (ICC) by one or three arbitrators in accordance with these Rules and shall be conducted in the ruling language of the Contract as given on Page 3 of the Contract Data.

The place of arbitration may be stated in the Contract Data. Otherwise, the prevailing convention is that the arbitration shall take place in a neutral country where appropriate expertise in international arbitration is available with access readily available to arbitrators with experience in the FIDIC Conditions of Contract. Preferably the chosen location should be equally physically accessible to the staff participants of both Parties.

In undertaking management of the arbitration hearing, the ICC will charge a fee commensurate with the scale and nature of the arbitration. Both Parties will require to pay in advance equal portions of these fees before the ICC will take any action.

From time to time, the ICC will demand additional payments as the arbitration continues. If a Party fails to pay its share of these additional fees, the other Party has two choices:

- it may pay the share of the defaulting Party, in which case the arbitration proceeding will continue. Credit will be given in respect of this additional payment in any final award.
- decline to pay the share of the defaulting Party, in which case the arbitration will be suspended or otherwise according to the ICC Rules.

The Arbitration has full power to open up, review, revise any certificate, determination (if not final and binding), determination opinion or valuation of the Certificate of the Engineer or any Decision of the Engineer. Nothing shall disqualify the Engineer from being called as a witness.

Prior to the opening of proceedings, legal advisors of both Parties will have identified those individuals they intend to call as witnesses. Each Party may cross-examine the witnesses called by the other Party. Any decision of the DAAB shall be admissible in evidence in the arbitration.

Arbitration may be commenced both before and after completion.

If an award requires a payment of an amount by one Party to the other Party, this amount shall be immediately due and payable without further certification or Notice.

The Conditions of Contract are silent on the possibility that the losing Party will not pay the amount of any award. Legal advice should be sought immediately.

21.7 Failure to Comply with DAAB's Decision

In the event that a Party fails to comply with any decision of the DAAB (binding or final and binding), then the other Party may refer the failure itself directly to arbitration under Sub-Clause 21.6 (Arbitration). The arbitral tribunal shall have the power to order the enforcement of that decision.

If the decision is binding but not final, the rights of the Parties are reserved until resolved by a final award.

Failure to comply with DAAB's decision may result in an award of damages.

21.8 No DAAB in Place

If a dispute arises between the Parties and there is no DAB in place (or being constituted), then

(a) Sub-Clause 21.4 (obtaining DAAB's Decision) and Sub-Clause 21.5 (Amicable Settlement) shall not apply and

(b) the Dispute may be referred by either Party directly to arbitration under Sub-Clause 21.6 (Arbitration) without prejudice to any other rights the Party may have.

Appendix A. Guidance for the Preparation of Particular Conditions

This book is focussed on the preparation of tenders and the award and execution of a contract in conformity with the standard 'General Conditions of Contract'.

However, these General Conditions require modification to comply with the Employer's specific requirements in respect of an individual project.

These specific requirements are addressed in the supplementary FIDIC document titled 'Guidance for the Preparation of the Particular Conditions'.

Two items require the attention of both Parties.

1. Particular Conditions Part A – Contract Data
   In the First Edition 1999, this document was titled 'Appendix to Tender'.
2. Particular Conditions Part B – Special Provisions
   In the First Edition 1999, this document was titled 'Particular Conditions of Contract'.

Contract Data

As part of the documentation provided to tenderers, the Employer will provide a listing derived from the Contract Data (Pages 3–7). The Employer may have added new items and deleted others. The Employer will have included data which can only be provided by him. This is important because the Contractor will need this information to complete his tender. The Contractor will complete the Contract listing in respect of items where only he has the necessary information. The Contractor's data input will be evaluated during the subsequent tender evaluation.

There are many references in the General Conditions of Contract to information and data to be provided in the Particular Conditions. Most of this information will be contained in the Contract Data as described above.

Guide to the FIDIC Conditions of Contract for Construction: The Red Book 2017, First Edition.
Michael D. Robinson.
© 2023 John Wiley & Sons Ltd. Published 2023 by John Wiley & Sons Ltd.

Special Provisions

The Employer may wish to modify selected clauses of the General Conditions of Contract for inclusion in the Special Provisions. Under the heading 'Notes on the Preparation of Special Provisions' FIDIC have provided *references and examples [showing] some of the Sub-Clauses in the General Conditions which may need amending to suit the needs of the project or the requirements of the project.*

It is also stated that

- Modifications shall suit the local law
- New wording must be suitable – reflecting existing definitions given in Clause 1 of the General Conditions.
- New wording should not change the obligations of the Parties.
- New wording should not create ambiguity or misunderstandings.

In their publication 'The FIDIC Golden Principles' (published in 2019), FIDIC expresses concern at the practices adopted by some employers.

There are five golden principles:

GP 1 – Duties, rights, obligations. . . must be generally as implied in the General Conditions

GP 2 – The Particular Conditions must be drafted clearly and unambiguously.

GP 3 – The Particular Conditions must not change the balance of risk-reward in the General Conditions

GP 4 – All time periods . . . must be of a reasonable duration.

GP 5 – . . . all formal disputes must be referred to the DAAB.

With reference to GP 1 above, FIDIC provides the following additional observations:

Example of proper application of GP 1:
The Employer is obliged to make payments under the Contract, irrespective of the Employer's financial arrangements.

Example of modifications that do not comply with GP 1:
Under a Red Book or Yellow Book contract, the Engineer is required to obtain the Employer's approval before making any determination of a Contractor's claim or granting any extension of time pursuant to Sub-Clause 3.7.

Appendix B. Employer's Claims

The Employer's claims identified in the Second Edition are similar to those contained in the First Edition, subject to minor changes in clause numbering:

1. Sub-Clause 1.13 (Compliance with Laws) provides for the Employer to claim additional costs should the Contractor fail to comply with his obligations under this sub-clause, thereby causing the Employer to incur additional costs.
2. Sub-Clause 4.19 (Temporary Utilities) requires the Contractor to provide all contemporary utilities, including electricity, gas, telecommunications, water, etc., which he requires for the execution of the Works. Should the Contractor be required to provide temporary utilities to the Employer and/or Engineer, this obligation may be included in the Particular Conditions or Specifications. Should the Employer already be established on site, he may choose to provide the temporary utilities (in whatever part). Terms and conditions may also be provided in the Particular Conditions or Specifications and clarify the fiscal responsibilities of the provider.
3. In the First Edition, Sub-Clause 4.20 (Employer's Equipment and Free-Issue Material) clarified the conditions by which any of these items would be included in the Works. The Employer shall, at his risk and cost, provide these materials at the time and place specified in the Contract. Additionally, the Employer remains responsible for any shortage, defect, or default, which may cause the Contractor to incur additional cost. This sub-clause has been deleted in its entirety in this Second Edition but may be re-introduced in the Particular Conditions or Specifications.

Employer's Claims under a CONS Contract

1.13   Compliance with Laws (Contractor fails to provide documentation)
7.5   Defects and Rejection (Defective Plant and Materials)
7.6   Remedial Works (Contractor fails to carry out remedial works)
8.7   Rate of Progress (Contractor adopts revised methods that cause the Employer additional costs)
8.8   Delay Damages (Contractor fails to complete on time)

Guide to the FIDIC Conditions of Contract for Construction: The Red Book 2017, First Edition.
Michael D. Robinson.
© 2023 John Wiley & Sons Ltd. Published 2023 by John Wiley & Sons Ltd.

9.2	Delayed Tests (Delay caused by Contractor)
9.4	Failure to Pass Tests on Completion (Only if Employer incurs additional costs)
11.3	Extension of Defects Notification Period (Due to Contractor's default)
11.4	Failure to Remedy Defects (Contractor fails to rectify)
11.8	Contractor to Search (Default is responsibility of Contractor)
11.11	Clearance of Site (Failure of Contractor to clear Site)
13.6	Adjustment for Changes in Laws (Decreases in cost only)
15.3	Valuation after Termination for Contractor's Defaults (Contractor's property valued for employer's benefit)
15.4	Payment after Termination for Contractor's Default (Employer may claim losses and damage after termination)
17.0	Indemnities (Employer claims of cost of events for which he is indemnified by Contractor)
19.1	General Requirements for Insurances (Employer makes claim if Contractor fails to insure)

The number of occurrences entitling the Employer to give a Notice of Claim is very similar to that stated in the First Edition. The Employer is now required to provide the Engineer with a Notice of Claim for each occurrence as it arises and follows the same procedures as the Contractor.

For several claims, the Employer is entitled to recover his costs. This entitlement would particularly apply to testing delayed by the Contractor or where he has failed to rectify defects in good order. Considering that the Engineer's staff are confirmed to be employees of the Employer, any potential additional costs may be related to additional work hours or standing time of the Engineer.

Appendix C. Contractor's Claims

Throughout this Second Edition the Contractor's claims and entitlements are described in greater detail than in the First Edition.

The following commentary identifies new or retitled Contractor entitlements for Costs and/or EOT together with details of any deletion of claim headings used in the First Edition.

Some sub-clause headings have been changed in the Second Edition. These are also noted to provide continuity for the benefit of personnel who are transiting from use of the First Edition to the Second Edition.

Sub-Clause 1.13 (Compliance with Laws) provides for the Contractor to claim additional costs consequent upon the failures of the Employer.

Sub-Clause 4.6 (Co-operation) shall, as stated in the Specification or instructed by the Engineer, allow opportunities for carrying out work by Employer's personnel, other contractors employed by the Employer (note, it does not state the other contractors are necessarily working on Site), and employees of Public authorities or private utility companies.

Sub-Clause 4.15 (Access Route)
This sub-clause has been modified. At the Base Date the Contractor is required to have satisfied himself of the availability and suitability of the access route(s) to the Site. Presumably, the Contractor will have stated his intended access route in his tender submittal. Should the designated access become unavailable because of changes required by the Employer or a third party after the Base Date, and a new access route must be provided, then the Contractor is entitled to claim his additional costs with EOT.

Sub-Clause 7.6 (Remedial Works)
Generally, the Contractor is responsible for any costs incurred in carrying out remedial works. This sub-clause has been slightly expanded and now confirms that the Contractor is entitled to reimbursement of his Costs (with EOT) if the remedial works arise because of

- any act by the Employer of the Employer's Personnel
- an Exceptional Event (Sub-Clause 18.4 refers)

Guide to the FIDIC Conditions of Contract for Construction: The Red Book 2017, First Edition.
Michael D. Robinson.
© 2023 John Wiley & Sons Ltd. Published 2023 by John Wiley & Sons Ltd.

<u>Sub-Clause 11.7 (Right of Access after Taking-Over)</u>
This sub-clause only occupied three lines of print in the First Edition and has been significantly expanded in the Second Edition.

The procedure to be followed by the Contractor to gain access to the Works is explained in detail. However, if access is unreasonably denied by the Employer, the Contractor shall be entitled to payment of any additional costs incurred.

This is a new claim heading.

<u>Sub-Clause 12.1 (Works to be Measured)</u>

<u>Sub-Clause 12.3 (Valuation of the Works)</u>
Should a disagreement arise concerning the amount of work measured or any valuation of the Works, the disagreement shall be referred to the Engineer to *'agree or determine'* the matter in accordance with Sub-Clause 3.7 (Agreement or Determination). Should the disagreement not be resolved, it will eventually become a formal claim to be processed in accordance with Sub-Clause 20.2 (Claims for Payment and/or EOT).

<u>Sub-Paragraph 13.3.2 (Variation by Request for Proposal)</u>
Should the Engineer request the Contractor to provide a formal proposal for a Variation, and the proposal is rejected or otherwise not taken action on, the Contractor is entitled to reimbursement of his Costs incurred in the preparation of the proposal.

This is a new claim heading.

<u>Sub-Clause 16.4 (Payment after Termination by Contractor)</u>
The Contractor's entitlements are set out in much greater detail than in the First Edition.

<u>Sub-Clause 17.2 (Liability for Care of Works)</u>
This sub-clause has been re-numbered. The Contractor's entitlements were originally contained in the First Edition Sub-Clause 17.4 (Consequences of Employer's Risks). The Contractor's entitlements are now described in greater detail.

<u>Sub-Clause 18.4 (Consequences of an Exceptional Event)</u>
This sub-clause has been re-numbered. The Contractor's entitlements were originally contained in the First Edition Sub-Clause 19.4 (Consequences of Force Majeure).

Contractor's Claims under a CONS Contract

Sub-Clause No.	Title	EOT
1.9	Delayed Drawings or Instructions	✓
1.13	Compliance with Laws	✓
2.1	Right of Access to the Site	✓
4.6	Co-operation	✓
4.7	Setting Out (Errors)	✓
4.12	Unforeseeable Physical Conditions	✓
4.15	Access Route	✓
4.23	Archaeological and Geological Findings	✓
7.4	Testing by the Contractor	✓

7.6	Remedial Works	✓
8.5	Extension of Time to Completion (General)	✓
8.10	Consequences of Employer's Suspension	✓
10.2	Taking Over Parts (Adjustment of Delay Damages)	X
10.3	Interference with Tests on Completion	✓
11.7	Right of Access after Taking Over	X
11.8	Contractor to Search	X
12.1	Works to be Measured (Disagreement referenced to Sub-Clause 3.7)	X
12.3	Valuation of the Works (Disagreement referenced to Sub-Clause 3.7)	X
13.3.2	Variation by Request for Proposal (abandoned by Employer)	X
13.6	Adjustments for Changes in Law (increases or decreases)	X
14.8	Delayed Payment (Financial Charges reference to Sub-Clause 20.2 not required)	X
14.4	Cessation of Employer's Liability (Final opportunity to make a claim)	X
15.6	Valuation after Termination for Employer's Convenience (refer to Sub-Clause 3.7)	X
16.1	Suspension by Contractor	✓
16.2	Termination (by Contractor)	✓
16.3	Contractor's Obligations after Termination	✓
16.4	Payment after Termination by Contractor	X
17.2	Liability for Care of Works (for rectification)	✓
18.4	Consequences of an Exceptional Event	✓
18.5	Optional Termination (by either Party)	X
20.2	Procedural Clause to be referenced by Contractor in making claim under the Contract	✓

Note:
✓ Indicates an EOT entitlement
X Indicates there is no EOT entitlement

For a majority of claims, the Contractor is entitled to payment of Cost together with EOT. For the remaining claims the Contractor's entitlement is limited to payment of Cost.

It is essential that the Contractor provides the Engineer with a Notice of claim for each occurrence entitling him to make a claim. Refer to Appendix D (Notices and Site Organisation).

Appendix D. Notices and Site Organisation

In the Notes to this Second Edition, it is stated
  '*. . . this edition provides*

(1) *greater detail and clarity on the requirements for notices and other communications,*
(2) *provisions to address Employer's and Contractor's claims equally and separated from disputes,*
(3) *mechanisms for dispute avoidance and*
(4) *detailed provisions for quality management, and verification of Contractor's contractual compliance*'.

In support of the above objectives, this Second Edition has introduced detailed requirements for the giving of Notices and other communications in Sub-Clause 1.3 (Notices and Other Communications).

'Notice' is defined in Sub-Paragraph 1.1.56 as '*means a written communication identified as a notice and issued in accordance with Sub-Clause 1.3*'. The number of occasions requiring a Party or the Engineer to give or respond to a Notice are extensive (refer to attached summary listing). All Notices must be given and responded to within strict time limits; otherwise, the originator of the Notice may lose any entitlements due to him.

Given the large number of Notices or other communications either to be given or responded to within strict time limits, both Parties and the Engineer must be well organised to deal with the ensuing administrative workload. It is a reasonable assumption that the bulk of the workload will be managed by the Engineer's Representative and the Contractor's Representative, since both are required to be permanently resident on Site.

Considering that communications will be made by electronic systems, it would be mutually beneficial if the two representatives were to agree an identification system which clearly identifies the purpose of any communication.

A typical heading of a communication could contain:

- Date of transmission (with auto-response).
- Recipient. . . . Copies to . . . .
- The title/subject of the transmission (e.g. "Notice of the Engineer's Decision")
- Code number/Contract Clause Reference

Guide to the FIDIC Conditions of Contract for Construction: The Red Book 2017, First Edition.
Michael D. Robinson.
© 2023 John Wiley & Sons Ltd. Published 2023 by John Wiley & Sons Ltd.

The respondent would respond with similar headings. This arrangement would not only minimise the possibility of misunderstandings but equally importantly would help provide a continuous record of any dispute which was unresolved for a lengthy period of time.

The Engineer's Representative and the Contractor's Representative are key figures in the organisation of their respective staff members. Both are likely to be signatories to any Notice or response and have ultimate responsibility for the content of the Notices. Consequently, the duties allocated to their respective staff members have to be carefully arranged to minimise the administrative burden on both the Engineer's Representative and the Contractor's Representative.

The attached organigram is typical of the site organisation of a medium -large construction project and shows the positions of not only the executives of the Employer, the Engineer, and the Contractor, but also the positions occupied by the senior on-site staff of the Engineer and the Contractor. Suggested lines of delegation are also shown.

A1. Delegation by the Engineer (to the Engineer's Representative on Site)
Sub-Clause 3.3 (The Engineer's Representative) states in part *'The Engineer may appoint an Engineer's Representative and delegate to him. . . the authority to act on the Engineer's behalf at the Site'.*

Note: This sub-clause does not mention any requirement for the Engineer to issue a Notice.

A2. Delegation by the Engineer (to assistants)
Sub-Clause 3.4 (Delegation by the Engineer) states *'The Engineer may from time to time assign duties and delegate authority to assistants. . . . By giving a Notice to the Parties, describing the assigned duties and the delegated authority of each assistant. . .'.*

Note:
This sub-clause requires the issuance of a Notice by the Engineer. The delegation of authority is made by the Engineer and not the Engineer's Representative. It is not specifically stated if Notices issued by the Contractor's Representative shall be directed to the Engineer or the Engineer's Representative. This requires clarification from the Engineer when issuing the delegation of duties to the Engineer's Representative.

B1. Delegation by the Contractor (to the Contractor's Representative)
The term "Contractor" in this context refers to an entity which has entered into the Contract with the Employer.

Sub-Clause 4.3 states in part *'The Contractor shall appoint the Contractor's Representative and shall give him all authority necessary to act on the Contractor's behalf under the Contract'.*

Note:
The appointment of the Contractor's Representative is subject to the approval of the engineer (not the Employer) and *'The Contractor's Representative may delegate any functions and authority except:*

- *the authority to issue and receive Notices and other communications. . .*
- *the authority to receive instructions under Sub-Clause 3.5 (Engineer's Instructions). . .'.*

Neither the Engineer's Representative nor the Contractor's Representative will be fully aware of the detail of each site activity. However, there are several key site activities which will require a Notice to be issued by one party and responded to by the other party with minimum delay.

Key activities include

1. Sub-Clause 7.3 (Inspection)

'The Contractor shall give a Notice to the Engineer whenever any materials, Plant or Work is ready for inspection, and before it is covered. . .' and 'The Employer's Personnel shall then either carry out the examination, inspection, measurement or testing without unreasonable delay, or the Engineer shall promptly give a Notice to the Contractor that the Employer's Personnel do not require to do so. . .'.

Typically, this could include inspections prior to concrete pours, covering of pipelines, etc. (but not testing of Plant). Appropriate authority could be delegated to the Contractor's senior field staff and laboratory and their opposite numbers from the Engineer's staff.

2. Sub-Clause 7.4 (Testing by the Contractor)

'The Contractor shall give a Notice to the Engineer, stating the time and place for the specified testing of any Plant, Materials and other parts of the Works'.

The bulk of site testing of concrete, soils, aggregates, rebar, etc., will be continuously carried out by the Contractor's site laboratory under the overall supervision of the equivalent Engineer's staff suitably delegated by the Engineer. Testing of complex items of Plant will be subject to a different regime and crucial testing will be the subject of a Notice from the Contractor to the Engineer.

3. Sub-Clause 4.7 (Setting Out) and Sub-Clause 12.1 (Works to be Measured)

Sub-Clause 4.7 (Setting Out) requires the contractor to verify the terms of reference, i.e. primary setting out of the project, executed by others and provided by the Employer. This activity is routinely performed by the Contractor, with the active participation of the Engineer's surveyors. Both parties will become aware of any errors in the provided survey data at the same time. The provision of Notice by the Contractor is a formality but should be observed, particularly if the Contractor is likely to suffer delay and/or incur cost.

4. Sub-Clause 12.1 (Works to be Measured) requires that *the Works shall be measured and valued for payment.* The clause continues *'Whenever the Engineer requires part of the Works to be measured on Site, he shall give Notice to the Contractor of not less than 7 days, of the part to be measured. . . '.*

(**Note**: There appears to be a grammatical error in this text. It is assumed that seven days is the minimum notice period.)

This sub-clause refers specifically to measurement on Site. A substantial portion of the work performed will be measured from drawings. Another portion will relate to excavations, filling operations, etc., which require the attendance of a surveyor before any work is hidden away or not otherwise accessible. This can be a continuous operation requiring the attendance of surveyors, consequently a seven days' notice period is untenable.

The surveyors of both parties should be delegated to agree any measurement as and when required. The most practical solution is that the Contractor's surveyor provides the Employer's surveyor with a Notice detailing the planned work requiring measurement (possibly on a weekly basis).

The issues of 'delegation' and 'Notices' require clarification by the Engineer and Contractor at a very early stage, preferably immediately following the award of the Contract.

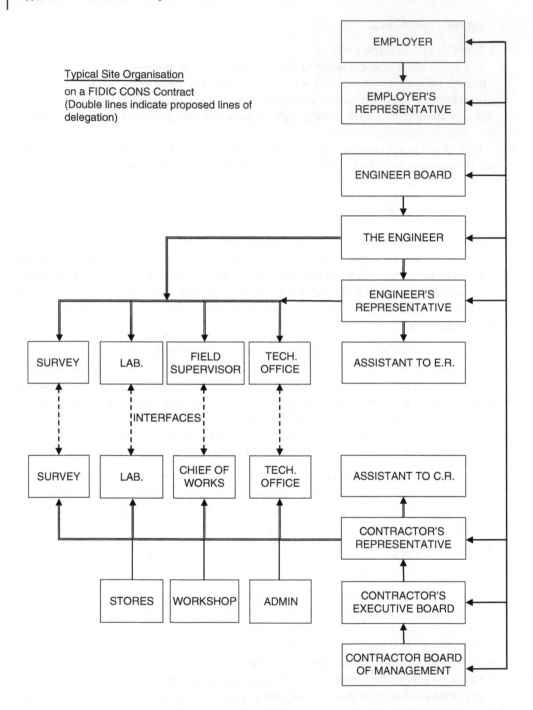

Typical Site Organisation

on a FIDIC CONS Contract
(Double lines indicate proposed lines of
delegation)

Listing of Notices identified in a CONS Contract

Clause/ Sub-Clause	Heading	Subject	Originator	Recipient
1.5	Priority of Documents	Ambiguity or Discrepancy	EMP/CONS	ENG
1.8	Care and Supply of	Error or Defect	All	All Documents
1.9	Delayed Drawing or Instructions	Delay to Works	CON	ENG
1.13	Compliance with Laws	Notice of Claim Sub-Clause 20.2	CON	ENG
2.1	Right of Access to Site	Notice of Claim Sub-Clause 20.2	CON	ENG
2.4	Employer's Financial Arrangements	Altered Financial Arrangements	EMP	CONS
3.1	The Engineer	Appointment of natural person acting on behalf of Engineer	ENG	EMP/CONS
3.3	The Engineer's Representative	Temporary Replacement of Engineer's Representative	ENG	CONS
3.4	Delegation by Engineer	Delegation by Engineer (1)	ENG	EMP/CONS
		Query of Delegation (2)	CONS	ENG
3.5	Engineer's Instructions	Objection to Engineer's Instructions	CONS	ENG
3.6	Replacement of Engineer	Replacement of Engineer (1)	EMP	CONS
		Temporary Delegation (2)	EMP	CONS
3.7.1	Consultation to reach Agreement	Engineer to give formal notice of agreement and if no agreement, he shall proceed to make Determination	ENG	EMP/CONS
3.7.2	Engineer's Determination	Engineer completes his Determination and issues 'Notice of Engineer's Determination'	ENG	EMP/CONS
3.7.3	Time Limits (of Agreements)	If Agreement is reached, the Engineer's Determination to be issued within 42 days	ENG	EMP/CONS
3.7.4	Effect of agreement or determination	Following Notice by a Party, errors to be corrected by Notice by Engineer	EMP/CONS	ENG
			ENG	EMP/CONS

(Continued)

Clause/ Sub-Clause	Heading	Subject	Originator	Recipient
3.7.5	Dissatisfaction with Engineer's	Dissatisfied Party to give Notice of Dissatisfaction (may relate to parts of the Engineer's Decision)	EMP/CONS	ENG Deter-mination
4.1	Contractor's General Obligations	Notice of No Objection to Contractor's Design before Commencement of Work	ENG	CONS
4.3	Contractor's Representative	Objection to appointment of CONS' Representative (1)	ENG	EMP/CONS
		Change of CONS' Representative (2)	CONS	ENG
4.4	Contractor's Documents	CON's documents ready for review (1)	CONS	ENG
		ENG no objection (2)	ENG	CNOS
4.7.2	Errors	Contractor to report any error in terms of reference	CONS	ENG
		Notice of Claim Sub-Clause 20.2	CONS	ENG
4.9.1	Q.M.	Engineer notifies Contractor of failures in Q.M.	ENG	CNOS
		Identified failing by CON	CONS	ENG
4.12.1	Unforeseeable Physical Conditions (U.P.C.)	Contractor to give Notice of U.P.C. to enable Engineer to investigate conditions	CONS	ENG
4.15	Access Route	Notice of Claim Sub-Clause 20.2	CONS	ENG
4.16	Transport of Goods	Contractor to give advance notice of arrival of Plant on Site	CONS	ENG
4.17	Contractor's Equipment	Contractor to give advance notice of arrival of Contractor's (main equipment) on Site	CONS	ENG
4.21	Security of Site	Visitors to be authorised by Notice	EMP/ENG	CONS
4.23	Archaeological Finds	Finding to be reported	CONS	ENG
5.1	Subcontractors	Contractor to obtain Engineer's consent (1)	CONS	ENG
		Engineer to give Notice of non-approval if necessary (2)	ENG	CONS
		Notice of commencement (3)	CONS	ENG
5.2.1	Nominated Subcontractors	Contractor objects to proposed subcontractor	CONS	ENG

Clause/ Sub-Clause	Heading	Subject	Originator	Recipient
5.2.4	Evidence of Payments	Engineer advises that Employer will pay subcontractors directly	ENG	CONS
6.12	Key Personnel	Contractor to obtain consent (1)	CONS	ENG
		Engineer to object by Notice (2)	ENG	CONS
7.3	Inspection	Contractor ready for inspection (1)	CONS	ENG
		Engineer to state if inspection is not required (2)	ENG	CONS
7.4	Testing by Contractor	Contractor to give Notice of time and place of testing (1)	CONS	ENG
		Engineer to give Notice of attendance (2)	ENG	CONS
		Notice of Claim Sub-Clause 20.2 (3)	CONS	ENG
7.5	Defects and Rejection	Engineer to give Notice of defects (1) and	ENG	CONS
		Second Notice of failure to correct defects (2)	ENG	CONS
		Employer costs – claim (3)	EMP	ENG
8.1	Commencement of Works	Notice of Commencement Date	ENG	CONS
8.3	Programme	Engineer gives Notice of modifications required (1)	ENG	CONS
		Engineer to give Notice of no objection (2)	ENG	CONS
		Engineer to give Notice of Failure to maintain programme of work (3)	ENG	CONS
8.5	Extension of Time for Completion	Notice of Claim Sub-Clause 20.2 (2)	CONS	ENG
8.7	Rate of Progress	Engineer to give Notice if Contractor proposals are not acceptable	ENG	CONS
8.8/8.10	Delay Damages	Employer to claim Delay Damages	EMP	ENG
8.12	Prolonged Suspension	After 84 days of stoppage Contractor gives Notice requesting recommencement	CONS	ENG

(Continued)

Clause/ Sub-Clause	Heading	Subject	Originator	Recipient
8.13	Resumption of Work	Engineer to give Notice to resume Work (1)	ENG	CONS
		Contractor to request re-start (2) Notice of Claim	CONS	ENG
		Sub-Clause 20.2 (3)	CONS	ENG
9.1	Contractor's Obligations	Engineer gives Notice that Contractor's proposals are non-compliant with Contract (2)	ENG	CONS
		Engineer gives No-objection (1)	ENG	CONS
9.2	Delayed Tests	If tests are delayed by Contractor, then Engineer can give Notice to commence testing within 21 days (1)	ENG	CONS
		If Contractor fails to recommence Work, Engineer will use own staff (2)	ENG	CONS
10.1	Taking Over of the Works and Sections	Contractor applies for Taking Over Certificate by Notice to Engineer (1)	CONS	ENG
		Engineer to issue Notice of Rejection (2)	ENG	CONS
10.2	Taking Over Parts	Contractor applies for Taking Over Certificate for Part taken into use (or agreed by Employer)	CONS	ENG
		Notice of Claim Sub-Clause 20.2	CONS	ENG
		Notice under Sub-Para 3.7.3	CONS	ENG
10.3	Interference with Tests	Delay caused by Employer's Personnel. Contractor to give Notice of prevention	CONS	ENG
		Engineer gives Notice to carry out tests	ENG	CONS
		Notice of Claim Sub-Clause 20.2	CONS	ENG
11.1	Completion of Outstanding Work and Remedying Defects	Contractor to execute all works required to remedy defects etc. (1)	ENG	CONS
		Contractor to respond to Reported defects (2)	ENG	CONS
11.2	Cost of Remedying Defects	Contractor to give Notice of any defects not attributable to him	CONS	ENG

Clause/ Sub-Clause	Heading	Subject	Originator	Recipient
11.3	Extension of DNP	Employer entitled to DNP plus costs	EMP	ENG
11.4	Failure to Remedy Defects	Employer to issue Notice of fixed date to complete repairs (1)	EMP	CONS
		Employer entitled to claim costs (2)	EMP	ENG
11.5	Remedying Defective Works Off Site	Contractor to give Notice of irreparable item as prelude to replacement	CONS	EMP
11.6	Further Tests after Remedying Defects	Within 7 days of completion of the Work Contractor to give a Notice providing proposals for remedying defects	CONS	ENG
		Failing which Engineer can give Notice for retesting	ENG	CONS
11.7	Right of Access after Taking Over	If Contractor requires access to Works during DNP, Notice to be given to Employer (1)	CONS	EMP
		Employer to give consent (2)	EMP	CONS
		Contractor to claim for unreasonable delay (3)	CONS	ENG
11.8	Contractor to Search	Employer fixes date of search	ENG	ENG
		Notice of Claim by Contractor	EMP	CONS
11.11	Clearance of Site	Employer to claim cost of clearing Site	EMP	CONS
12.1	Works to be Measured	Engineer to give 7 days' Notice of date for measurement of the Works	ENG	CONS
		Contractor considers measurement is incorrect	CONS	ENG
12.3	Valuation of the Works	Contractor to give Notice of disagreement of Engineer's measurement (1)	CONS	ENG
		Contractor to give Notice of disagreement with Engineer's measurement (2)	CONS	ENG
12.4	Omissions	Contractor to give Notice of unreimbursed costs	CONS	ENG
13.1	Right to Vary	Contractor to give Notice of any inability to execute a given Variation (1)	ENG	CONS
		Engineer to respond (2)	ENG	CONS

(Continued)

Clause/ Sub-Clause	Heading	Subject	Originator	Recipient
13.2	Value Engineering	Engineer to respond to Contractor's Value Engineering proposals	ENG	CONS
13.3.1	Variation by Instruction	Engineer may instruct variation	ENG	CONS
		Notice of Claim Sub-Clause 20.2	CONS	ENG
13.3.2	Variation by Request for Proposal	Engineer to initiate request for Variation (1)	ENG	CONS
		Contractor to respond (2)	CONS	ENG
13.4	Provisional Sums	Contractor to provide Quotations Engineer to select	ENG	CONS
13.6	Adjustment for Changes in Laws	Adjustment to the execution of the Works (1)	CONS	ENG
		Engineer to respond (2)	ENG	CONS
14.4	Schedule of Payments	Engineer to determine revised instalments	ENG	EMP/CONS
14.6.2	Withholding of an IPC	Amount of IPC less than stated minimum Notice required	ENG	CONS
14.6.3	Corrections or Modification	Contractor objects to correction/modification to IPC	CONS	ENG
		Notice of Claim	CONS	ENG
14.11.1	Final Statement	Engineer disagrees with Contractor's Draft Final Statement	ENG	CONS
14.11.2	Agreed Final Statement	Contractor to submit	CONS	ENG
14.14	Cessation of Employer's Liability	Final Claims (if any)	CONS	ENG
15.1	Notice to Correct	Contractor fails to carry out obligations	ENG	CONS
		Contractor to provide Remedial measure proposals	CONS	ENG
15.2.1	Termination for Contractor's Default	Employer's intention to terminate due to Contractor's Failure	EMP	CONS
15.2.2	Termination	Should Contractor continue to default, Employer may terminate (14 days' notice) (2)	EMP	CONS

Clause/ Sub-Clause	Heading	Subject	Originator	Recipient
15.2.4	Completion of Works by Employer	Use of Contractor's Equipment to complete Works	EMP	CONS
		Instructions for collection on Completion	EMP	CONS
15.5	Termination for Convenience	Termination for Employer's convenience	EMP	CONS
15.6	Valuation after Termination for Employer's Convenience	Contractor to submit valuation (1)	CONS	ENG
		Engineer to make Determination (2)	ENG	CONS
16.1	Suspension by Contractor	Failure of Employer Suspension by Contractor	CONS	EMP
		Notice of Claim by Contractor	CONS	ENG
16.2.1	Termination by Contractor (Notice)	Various failures of Employer leading to termination by Contractor unless remedied	CONS	EMP
16.2.2	Termination	Employer fails to remedy within 14 days. Second Notice of Termination	CONS	EMP
		Notice of claim by Contractor	CONS	ENG
17.2	Liability for Care of the Works	Exceptional event (Clause 18) causes damage to Works	CONS	ENG
		Notice of claim by Contractor	CONS	ENG
18.2	Notice of Exceptional Event	Exceptional Event Prevention of Work	CONS or EMP	EMP or CONS
18.3	Duty to Minimise Delay	Exceptional Event ends Work recommences	CONS or EMP	EMP or CONS
18.5	Optional Termination	After 84 days' delay (single event) or 140 days (multiple days) either Party may give Notice of Termination	CONS or EMP	EMP or CONS
		Engineer to make Determination	ENG	EMP and CONS
18.6	Release from Performance	No agreement to permit continued performance of Works	CONS	EMP c.c. ENG
20.1	Claims	Parties entitled to raise claims for other matters not specifically stated in Contract	ENG or CONS	CONS or ENG

(Continued)

Clause/ Sub-Clause	Heading	Subject	Originator	Recipient
20.2.1	Claims for Payment and/or EOT	Claiming Party to give Notice to Engineer no later than 28 days of claiming Party becoming aware of event	CONS or EMP	ENG
20.2.2	Engineer's initial response	Should the Engineer consider that the time limits specified in Sub-Paragraph 20.2.1 are exceeded, he shall give a Notice to the claiming Party with his reasons	ENG	CONS or EMP
20.2.4	Fully detailed Claim	(1) Claiming Party to submit fully detailed claim within 84 days of event, otherwise Claim will lapse	CONS	ENG
		(2) Notice of claim lapse (Subject to 14 days lapse provision)	ENG	CONS or EMP
20.2.5	Agreement or Determination of the Claim	(1) Resolution of validity of submittal of notices	ENG	CONS or EMP
		(2) Engineer requires further particulars	ENG	CONS or EMP
		(3) Engineer gives response on legal validity of claim	ENG	CONS or EMP
21.4.1	Reference of Dispute to DAAB	A Party is dissatisfied with Engineer's NOD (Sub-Clause 3.7) to be referred to DAAB within 42 days of date of NOD	EMP or CONS	ENG
21.4.3	DAAB Decision	Decision within 84 days of Referral	DAAB	All
21.4.4	Dissatisfaction with DAAB decision	NOD within 28 days of receiving DAAB decision	EMP or CONS	DAAB

Appendix E. Daywork and Contemporary Record Sheets

(1) Dayworks

The opening sentence of Sub-Clause 13.5 (Dayworks) states *'If a Daywork Schedule is not included in the Contract, this Sub-Clause shall not apply'*. If there is no Daywork Schedule included in the tender documents, then payment cannot be made on a Daywork basis. Sub-Clause 12.3 (Valuation of the Works) provides opportunity for valuation of Dayworks to be negotiated, provided they are of a minor and incidental nature.

The supplementary documents 'Guidance for the Preparation of Particular Conditions' and 'Notes on the Preparation of Special Provisions' do not provide any indicators of what requirements would be included in a Daywork Schedule.

One solution would be for the inclusion in the Tender Documents of a 'Schedule of Dayworks' prepared by the Employer with a limited number of items most relevant to Daywork operations. The quantities of each of the Schedule items are also to be provided by the Employer. The Schedule shall be priced by Tenderers, noting that the unit prices are deemed to include taxes, overhead and profit. The total value of the Schedule is to be included as a Provisional Sum in the Bill of Quantities. This procedure ensures that the daywork rates offered by each tenderer is reflected in the total tendered amount.

The procedure given in Sub-Clause 13.5 (Dayworks) for obtaining any Goods for use on a Daywork task is cumbersome. To avoid delays and reduce costs, much of the required Goods are frequently sourced from the Contractor's own stores (where supporting invoices are available). The Engineer is authorised to instruct the Contractor to obtain other quotations (for Goods not immediately available with the Contractor's stores) from other sources.

Sub-Clause 13.5 (Dayworks) requires the Contractor *'to deliver each day to the Engineer accurate statements in duplicate. . . which shall include records (as described in Sub-Clause 6.10 (Contractor's Records)) of the resources used in executing the previous day's work'*. These records are to be priced for inclusion in the next Statement.

It is noted that the rates and prices in the daywork Schedule (if included in the Contract) shall be deemed to include taxes, overhead, and profit.

Guide to the FIDIC Conditions of Contract for Construction: The Red Book 2017, First Edition.
Michael D. Robinson.
© 2023 John Wiley & Sons Ltd. Published 2023 by John Wiley & Sons Ltd.

Following is a proposed standard format which can be varied to suit the requirements of an individual site. These daywork/daily record sheets should be colour-printed in groups of five and bound in sets of 100 (20 × 5). Each set of five would have a unique sequential number. This would help identify records that are not signed, not presented, or simply 'lost' in the system. The top copy and one other would be taken by the Engineer and the next two copies by the Contractor's site office. These four copies would be perforated for separation. A fifth copy, unperforated, would be kept in the Contractor's field office responsible for producing the record sheets. It may happen that the Contractor's supervisory staff are not fluent in the language of the Contract. Provided that Badge No./Fleet No./Material Codes are correctly recorded, together with the relevant quantities, the form can be conveniently completed by a junior staff member with the correct language skills.

The valuation of the individual sheets takes place in the Contractor's site office. The quantum, whether for daywork or other purposes, can be sent to the Engineer as part of the monthly Application for Interim Certificate or as part of a claim presentation accompanied by a copy of the original, signed daywork/daily record sheet.

2) Contemporary Records

Sub-Clause 20.2 (Claims for Payment and/or EOT) states in part:

'The claiming Party shall keep contemporary records as may be necessary to substantiate the claim'

and

'The Engineer may monitor the Contractor's contemporary records and/or instruct the Contractor to keep additional contemporary records'.

Routinely, the Contractor generates a considerable amount of data for his own purposes (cost control, taxes, financial planning, etc.). This can be made available to the Engineer for his review (subject to confidentiality). However, the evaluation of some claims will require more specific records of actual site events relating to the basis of any claim. These may require the Contractor to maintain contemporary records in more detail than for dayworks. Diaries, minutes of meetings, correspondence, construction drawings, etc., will all form part of Contemporary Records.

Any requirement for 'Contemporary Records' applies to any claim made by the Employer and which is to be evaluated by the Engineer.

Date   DAYWORK/DAILY RECORD SHEET   Sheet Reference

....

PROJECT .... SECTION ....

DESCRIPTION ....

REFERENCE (if any) ....

	Badge	Name	Trade	Hours		Fleet No.	Description	Hours		Code	Description	Quantity	Unit
LABOUR					**EQUIPMENT**				**METERIALS**				
1					1				1				
2					2				2				
3					3				3				
4					4				4				
5					5				5				
6					6				6				
7					7				7				
8					8				8				
9					9				9				
10					10				10				
11					11				11				

CONTRACTOR: .... ENGINEER: ....

Distribution: Engineer: white + yellow
                 Contractor's Office: green + pink
                 Site: blue – fast copy

Appendix F. Contractor's Costs

(1) The term 'Cost' is defined in Sub-Paragraph 1.1.9: Cost means all expenditure reasonably incurred (or to be incurred) by the Contractor in performing the Contract, whether on or off the Site, including taxes, overheads and similar charges.

(2) 'Overheads' may be more broadly defined as those expenses that together with the taxes and planned profit are not directly chargeable to the cost of production. Production costs, principally the cost of the plant, labour, and materials are conventionally referred to as direct costs. Thus, the summation of overhead costs and direct costs represents the total cost of a given project to the Contractor. Additionally, a percentage for profit and risk is to be added to the total cost to give the Contractor's tender price. Occasionally, in other literature, the reader may find reference to 'indirect costs' in substitution to 'overhead costs'. For consistency, this book follows the FIDIC preference for 'overhead'.

(3) In order to maintain commercial confidentiality, access to the Contractor's tender summary sheet or so-called 'top sheet' is normally restricted to a limited number of the Contractor's senior management.

   Nonetheless, it seems inevitable that at some point essential pricing data will have to be made available to the Engineer in order to facilitate the valuation of variations, claim valuation, and settlement, particularly those claims related to extension of time.

   It has been the author's experience that it is preferable that the principal features of the Contractor's 'top sheet' are presented to, and accepted by, the Employer/Engineer at an early stage of the Contract and before claims and disputes arise. It becomes increasingly difficult to reach a general agreement on the value of overheads after claims or disputes have arisen, particularly if the Engineer's relevant site staff have little understanding of the principles used in the compilation of the Contractor's 'top sheet'.

   The contractor's 'top sheet' is prepared by his estimating office and requires adjustment prior to commencement of the Works to reflect the consequences of agreements reached in any post-tender meetings with the Employer.

(4) Conventionally, the Contractor's overhead costs are expressed as a percentage of direct costs even though the Contractor's estimating office will, for the most part, have calculated the various elements of the overhead costs as individual sums of money. This

Guide to the FIDIC Conditions of Contract for Construction: The Red Book 2017, First Edition.
Michael D. Robinson.
© 2023 John Wiley & Sons Ltd. Published 2023 by John Wiley & Sons Ltd.

percentage can vary considerably from more than 40% on a large, isolated project (where the Contractor has to provide a large number of facilities and services, e.g. housing, schools, hospitals) to less than 20% on a small project in a developed location, where site establishment is minimal and full use is made of local suppliers and services.

The Contractor's overhead costs may be reduced if some of these costs are paid directly by means of specific bill items in the Bill of Quantities.

A proportion of the overhead will be incurred off site ('Head Office Expenses') and the remainder on the site ('Site Office Expenses'). Both the Head Office Expenses and the Site Office Expenses can be sub-divided into time-related costs and fixed-overhead costs.

The Contractor's 'top sheet' will typically contain the following information:

Direct Costs	%	%
Equipment	30	
Labour	25	
Materials	25	
Sub-contractors	<u>20</u>	100
Head Office Expenses		
Time-Related Overhead Costs	4	
Fixed-Overhead Costs	<u>2</u>	6
Site Office Expenses		
Time-Related Overhead Costs		
• expatriate staff		
• non-productive local staff		
• office vehicles		
• consumables		
• communications	9	
Fixed-Overhead Costs		
• provide office, etc.	5	
• establishment demobilisation costs	<u>2</u>	16
Total Costs		122
Profit, Risk, Financing (6% of total cost)	<u>7.32</u>	<u>7.32</u>
		<u>129.32</u>

The breakdown of the Accepted Contract Amount demonstrated above is an important aid in the following matters:

(a) the agreement of new rates for changes in quantities of work to be performed (Sub-Clause 12.3 refers)

(b) the agreement of new rates for works that are the subject of a Variation (Sub-Clause 12.3 refers)

(c) the evaluation of financial claims arising from any award of an Extension of Time for Completion (Sub-Clause 8.5 refers). As summarized in Appendix A, there are 14 in number of events or circumstances that permit the Contractor to claim additional payment in respect of Extension of Time for Completion. Of those fourteen instances, four restrict the Contractor's entitlement to his costs only.

In the administration of those matters summarised above, there is a general assumption that Overhead Costs are spread evenly over all bill items and over the full period of the Contract. In particular, Variations made under the provisions of Sub-Clause 12.3 are most likely to be evaluated on this assumption. It is impractical to assess the individual overhead applicable to individual bill items, and the use of a global average percentage for all items is seemingly inevitable, excluding only those items where specific alternative measures are provided (e.g. the percentage for adjustment of Provisional Sums – Sub-Clause 13.4 refers).

Referring to the breakdown given in the previous page, it can be stated as a generalisation that a fixed overhead of 29.32% would be applicable to all new evaluations made under the provisions summarised in (a), (b), and (c) above, excluding items where the evaluation is specified to be based on cost only.

Nonetheless, there are occasions when the practice of averaging overheads over the period of the Contract is not appropriate. The reality is that the Contractor does not incur his Overhead Costs evenly over the period of the Contract.

Mobilisation Period

In the initial stages of a Contract, it takes time for the Contractor to mobilise his resources and establish himself on Site. His expenditure on Fixed-Overhead Costs will be relatively high, and Time-Related Costs will gradually increase over this period as more and more operations are commenced.

Construction Period

Once the mobilisation period is completed, expenditure on Fixed Overheads can be assumed to fall to a relatively constant level for the construction period. Equally, Time-Related Costs can also be assumed to fall to a relatively constant level, although this depends considerably on the incidence of high-value work items. These Time-Related Costs will eventually reduce as the date of Taking Over approaches, when sections of the Works are completed, and the Contractor's resources are reduced.

Defects Liability Period

Fixed-overhead Costs will chiefly derive from the final demobilisation of the Contractor's resources. Time-Related Costs will relate to completion of outstanding works and the correction of defects, which under average circumstances will be largely complete within the first few months of the Defects Liability Period.

If no specific cost data is available, the incidence of expenditure for all Overhead Costs may be estimated from other documents, including the Contractor's cash-flow projections and the Programme of Works.

Appendix G. Joint Ventures

(a) <u>Full Joint Venture Agreement for the Contractor's internal use only</u>
Individual contractors may decide that a project is too large or too complex, with attendant high risk, to be executed solely on their own account, and consequently they will seek reliable partners to reduce the risk to an acceptable level.

Many contractors will have worked together on other projects and would be prepared in a joint venture for another project.

Having decided to form a Joint Venture for a specific project, a Full Joint Venture Agreement will be drawn up, which will precisely define the many elements of the Joint Venture. This document should be drawn up as soon as possible.

Principal elements to be considered are listed below:

 (1) Definitions
 (2) Joint Venture (Structure and Name)
 (3) Commencement and Term
 (4) Preparation and Submission of Tender
 (5) Failure to agree Tender or Unsuccessful Tender
 (6) Successful Tender
 (7) Profits and Losses
 (8) Tender Costs
 (9) Supervisory Board
(10) Executive Board
(11) Sponsor
(12) Project Manager
(13) Liability of the Joint Venturers to each other
(14) Bank Accounts
(15) Working Capital
(16) Invoicing
(17) Partners' Accounts
(18) Provision of Resources and Services
(19) Realization of Assets
(20) Information
(21) Security

Guide to the FIDIC Conditions of Contract for Construction: The Red Book 2017, First Edition.
Michael D. Robinson.
© 2023 John Wiley & Sons Ltd. Published 2023 by John Wiley & Sons Ltd.

(22) Insurance

(23) Taxation

(24) Confidentiality

(25) Exclusivity

(26) Assignment

(27) Publicity

(28) Default

(29) Nature of Agreement

(30) Illegality

(31) Amendments

(32) Notices

(33) Waivers

(34) Applicable Law

(35) Arbitration

(36) Duration of Agreement

(b) Joint Venture Prequalification (if required)

The prequalification process, initiated by the Employer, typically by means of advertising, will describe how to obtain prequalification documents containing all necessary instructions which are to be followed by prospective tenderers. The submittal of prequalification documents will require individual contractors having to decide whether to prequalify solely on their own account or as part of a joint venture or consortium.

Having decided to join with other contractors in a Joint Venture, the Contractor is effectively committed not only to involvement in the prequalification and tender process, but also to participate in the execution of the project (if awarded).

(c) Joint Venture Agreement (JVA) for submittal to the Employer

This is a condensed version of the Full Joint Venture Agreement discussed in (a) above, with confidential matter removed.

Immediately following an award of the Contract to the JV, this document is to be prepared and submitted to the Employer for approval and inclusion in the Contract documents.

Model Form of a Pre-Bid Joint Venture Agreement

This Agreement is made and entered into this (date)

By and between

Contractor 'A', address, hereinafter referred to as . . . . . .

and

Contractor 'B', address, hereinafter referred to as . . . . . .

The above named may also be referred to individually as 'Party' and collectively as 'Parties'.

WHEREAS

A   (Describe title and address of the prospective Employer), hereinafter referred to as the 'Employer', has invited tenders (the 'Tender') for the construction of (describe the project), hereinafter referred to as the 'Project' or 'Works'

B   The Parties wish to associate themselves in a joint venture (hereinafter referred to as the 'Joint Venture') for the purpose of submitting the Tender and if the Tender is successful, to enter into a contract (hereinafter referred to as the 'Contract') to jointly execute the Works.

Now therefore it is hereby agreed as follows:

1. The parties do hereby form the Joint Venture and agree to jointly prepare and submit the tender with the express intention to jointly performing the Contract should be Employer accept the Tender. For these purposes, the Parties are jointly and severally bound towards the Employer and/or third parties (see Note following).

Note:
The model form provided assumes that all joint venture partners will jointly submit the tender by signing the relevant parts of the tender documents. Each signatory will be required to provide documentary of his authority (power of attorney) to submit the tender. In some instances (e.g. when the partners are resident in different countries), it is recommended for the senior partner to be authorised to sign and submit the tender on behalf of all the joint venture partners. If this procedure is to be followed, it would

Guide to the FIDIC Conditions of Contract for Construction: The Red Book 2017, First Edition.
Michael D. Robinson.
© 2023 John Wiley & Sons Ltd. Published 2023 by John Wiley & Sons Ltd.

be appropriate for the senior partner to be given the title and duties of Sponsor. The role of the Sponsor is given in Chapter 4 following.

2. The Joint Venture shall be known as . . . . or by such other name as hereafter the Partners agree.
3. The percentage of participation of the Parties in the Joint Venture shall be as follows:
   Contractor 'A'. . .% (describe those parts of the tender to be prepared)
   Contractor 'B'. . .% (describe ditto as above)
      Each Party shall furnish in accordance with the above percentages its respective share in any deposit, bond and other security required in connection with the Tender.
4. The preparation of the Tender shall be shared between the Parties as far as practical in the participation shares given in the item above. Each Party shall bear all and any costs they may occur in the participation of the Tender (see Note 4.4 of attachment).
5. The Tender shall be jointly submitted by the Parties. Each Party shall provide all appropriate authorisations and documentation that will permit the Tender to be submitted in compliance with the Tender requirements specified by the Employer.
6. Not Party hereto shall sell, assign, or in any manner encumber or transfer its interest or any part thereof in this Agreement or the Tender, whether by operation of the law or otherwise, without obtaining the consent of the other Party.
7. Except through this Joint Venture, no Party to this Agreement shall directly tender for or take interest for its own benefit in tendering for or in the execution of the Project.
8. The Joint Venture shall cease to exist, and this Agreement shall terminate in the event that the Contract has not been awarded to the Joint Venture until (date) (as such date may be extended by the Parties in writing) and the Tender surety (if any) has been returned to the Employer.
9. The Parties hereby agree that in the event of the Tender being accepted by the Employer, this Pre-Bid Joint Venture Agreement shall be superseded by a Joint Venture Agreement, which shall fully detail the legal framework of the Joint Venture and the responsibilities and duties of the respective Joint Venture partners.
10. This Agreement shall be governed by and construed in accordance with the laws of . . . . (specify).

Note:
It is important that the law governing this Pre-Bid Joint Venture be clearly specified. The Partners may be of differing nationalities and laws of a country perceived to be "neutral" may be preferred. The laws of Switzerland are a common choice for European-based contractors and are likely to be acceptable to any local (i.e. based in the country of execution) partner. Legal advice may be required.

11. The Parties shall endeavour to settle amicable any dispute arising out of or in connection with this Agreement. Failing amicable settlement, any such dispute shall be finally settled under the Rules of Arbitration of the International Chamber of Commerce, Paris, by a single arbitrator appointed in accordance with the said Rules. The venue of arbitration shall be . . . (specify place and country) and the proceedings shall be conducted in the . . . (specify) language.

    Judgement upon the award rendered may be entered in any court having jurisdiction or application may be made to such court for a judicial acceptance or the award and an order of enforcement, as the case may be.

    No Party shall be released from due performance under the terms and conditions of this Agreement because an arbitral proceeding has been initiated.

Note:

The Rules of Arbitration in this model are stated to be those of the International Chamber of Commerce, Paris. Other rules are available, such as UNCITRAL.

The venue for any arbitral proceedings can be any place mutually convenient to the partners.

The proceedings can be conducted in any language, but there is merit in selecting the language of the construction contract.

These choices are likely to be reflected in the adjudication of disputes with the Employer under the construction contract.

IN WITNESS WHEROF, the Parties execute this Agreement the day and year fist above written.

CONTRACTOR "A"   CONTRACTOR "B"

(date)   (date)

Index of Clauses and Sub-Clauses

(Sorted according to FIDIC Clause numbering system)

Clause		Page
1	**General Provisions**	
1.1	Definitions	1
1.2	Interpretation	6
1.3	Notices and Other Communications	6
1.4	Language and Law	7
1.5	Priority of Documents	8
1.6	Contract Agreement	8
1.7	Assignment	9
1.8	Care and Supply of Documents	9
1.9	Delayed Drawings or Instructions	9
1.10	Employer's Use of Contractor's Documents	11
1.11	Contractor's Use of Employer's Documents	11
1.12	Confidentiality	11
1.13	Compliance with Laws	12
1.14	Joint and Several Liability	12
1.15	Limitation of Liability	13
1.16	Contract Termination	13
2	**The Employer**	
2.1	Right of Access to the Site	15
2.2	Assistance	16
2.3	Employer's Personnel and Other Contractors	17
2.4	Employer's Financial Arrangements	17
2.5	Site Data and Terms of Reference	17
2.6	Employer-Supplied Materials and Employer's Equipment	18

Clause		Page
3	**The Engineer**	
3.1	The Engineer	19
3.2	Engineer's Duties and Authority	19
3.3	The Engineer's Representative	20
3.4	Delegation by the Engineer	20
3.5	Engineer's Instructions	21
3.6	Replacement of the Engineer	21
3.7	Agreement or Determination	21
3.8	Meetings	24
4	**The Contractor**	
4.1	Contractor's General Obligations	27
4.2	Performance Security	27
4.3	Contractor's Representative	29
4.4	Contractor's Documents	29
4.5	Training	31
4.6	Cooperation	31
4.7	Setting Out	32
4.8	Health and Safety Obligations	32
4.9	Quality Management and Compliance Verification	33
4.10	Use of Site Data	34
4.11	Sufficiency of the Accepted Contract Amount	35
4.12	Unforeseeable Physical Conditions	35
4.13	Rights of Way and Facilities	38
4.14	Avoidance of Interference	38
4.15	Access Route	38
4.16	Transport of Goods	39
4.17	Contractor's Equipment	40

Guide to the FIDIC Conditions of Contract for Construction: The Red Book 2017, First Edition.
Michael D. Robinson.
© 2023 John Wiley & Sons Ltd. Published 2023 by John Wiley & Sons Ltd.

Clause		Page
4.18	Protection of the Environment	40
4.19	Temporary Utilities	40
4.20	Progress Reports	41
4.21	Security of the Site	42
4.22	Contractor's Operations on Site	42
4.23	Archaeological and Geological Findings	43
5	**Subcontracting**	
5.1	Subcontractors	45
5.2	Nominated Subcontractors	46
6	**Staff and Labour**	
6.1	Engagement of Staff and Labour	49
6.2	Rate of Wages and Conditions of Labour	49
6.3	Recruitment of Persons	49
6.4	Labour Laws	50
6.5	Working Hours	50
6.6	Facilities for Staff and Labour	50
6.7	Health and Safety for Personnel	50
6.8	Contractor's Superintendence	51
6.9	Contractor's Personnel	51
6.10	Contractor's Records	52
6.11	Disorderly Conduct	52
6.12	Key Personnel	52
7	**Plant, Materials and Workmanship**	
7.1	Manner of Execution	3
7.2	Samples	53
7.3	Inspections (by the Employer's Personnel)	54
7.4	Testing by the Contractor (on Site)	55
7.5	Defection and Rejection	55
7.6	Remedial Work	56
7.7	Ownership of Plant and Materials	56
7.8	Royalties	57
8	**Commencement, Delays and Suspension**	
8.1	Commencement of Works	59
8.2	Time for Completion	59
8.3	Programme	60
8.4	Advance Warning	60

Clause		Page
8.5	Extension of Time	61
8.6	Delays caused by Authorities	62
8.7	Rate of Progress	62
8.8	Delay Damages	63
8.9	Employer's Suspension	64
8.10	Consequences of Employer's Suspension	64
8.11	Payment for Plant and Materials after Employer's Suspension	65
8.12	Prolonged Suspension	66
8.13	Resumption of Work	66
9	**Tests of Completion**	
9.1	Contractor's Obligations	69
9.2	Delayed Tests	70
9.3	Re-testing	70
9.4	Failure to Pass Tests on Completion	70
10	**Employer's Taking Over**	
10.1	Taking Over of the Works and Sections	71
10.2	Taking Over of Parts of the Works	72
10.3	Interference with the Tests on Completion	73
10.4	Surface Requiring Reinstatement	74
11	**Defects after Taking Over**	
11.1	Completion of Outstanding Work and Remedying Defects	75
11.2	Cost of Remedying Defects	76
11.3	Extension of Defects Notification Period	77
11.4	Failure to Remedy Defects	77
11.5	Remedying of Defective Work Off Site	77
11.6	Further Tests After Remedying Defects	78
11.7	Right of Access After Taking Over	78
11.8	Contractor to Search	78
11.9	Performance Certificate	79
11.10	Unfulfilled Obligations	79
11.11	Clearance of Site	79
12	**Measurement and Valuation**	
12.1	Works to Be Measured	81
12.2	Method of Measurement	83

Clause		Page	Clause		Page
12.3	Valuation of the Works	84	15.6	Valuation After Termination for Employer's Convenience	114
12.4	Omissions	85	15.7	Payment After Termination for Employer's Convenience	114
13	**Variations and Adjustments**				
13.1	Right to Vary	87	**16**	**Suspension and Termination by the Contractor**	
13.2	Value Engineering	89			
13.3	Variation Procedure	90	16.1	Suspension by the Contractor	115
13.4	Provisional Sums	91	16.2	Termination by the Contractor	116
13.5	Daywork	92	16.3	Contractor's Obligations After Termination	117
13.6	Adjustments for Changes in Law	93			
13.7	Adjustments for Changes in Cost	94	16.4	Payment After Termination by the Contractor	117
14	**Price and Payment**		**17**	**Care of the Works and Indemnities**	
14.1	The Contract Price	97			
14.2	Advanced Payment	97	17.1	Responsibility for Care of the Works	119
14.3	Application for Interim Payment Certificates	99	17.2	Liability for Care of the Works	119
14.4	Schedule of Payments	100	17.3	Intellectual and Industrial Property Rights	121
14.5	Plant and Materials Intended for the Works	101	17.4	Indemnities by Contractor	121
14.6	Issue of IPC	102	17.5	Indemnities by Employer	122
14.7	Payment	103	17.6	Shared Indemnities	122
14.8	Delayed Payment	103	**18**	**Exceptional Events**	
14.9	Release of Retention Money	104	18.1	Exceptional Events	123
14.10	Statement on Completion	105	18.2	Notice of an Exceptional Event	124
14.11	Final Statement	105	18.3	Duty to Minimise Delay	124
14.12	Discharge	106	18.4	Consequences of an Exceptional Event	124
14.13	Issue of FPC (Final Payment Certificate)	107	18.5	Optional Termination	125
14.14	Cessation of Employer's Liability	107	18.6	Release for Performance Under the Law	126
14.15	Currencies of Payment	107	**19**	**Insurance**	
15	**Termination by the Employer**		19.1	General Requirements	127
			19.2	Insurance to be Provided by the Contractor	129
15.1	Notice to Correct	109			
15.2	Termination for Contractor's Default	110	**20**	**Employer's and Contractor's Claims**	
15.3	Valuation After Termination for Contractor's Default	112	20.1	Claims	134
			20.2	Claims for Payment and EOT	134
15.4	Payment After Termination for Contractor's Default	113	**21**	**Disputes and Arbitration**	
			21.1	Constitution of the DAAB	139
15.5	Termination for Employer's Convenience	113	21.2	Failure to Appoint DAAB Members	140

Clause	Page	Clause	Page
21.3 Avoidance of Disputes	140	Appendix B. Employer's Claims	147
21.4 Obtaining DAAB's Decision	140	Appendix C. Contractor's Claims	149
21.5 Amicable Settlement	143	Appendix D. Notices and Site Organisation	153
21.6 Arbitration	143		
21.7 Failure to Comply with DAAB's Decision	144	Appendix E. Daywork and Contemporary Record Sheets	165
21.8 No DAAB in Place	144		
Appendices		Appendix F. Contractor's Costs	169
Appendix A. Guidance for the Preparation of Particular Conditions	145	Appendix G. Joint Ventures	173

Index of Sub-Clauses

Based on the original FIDIC document

	Sub-Clause	Page		Sub-Clause	Page
Accepted Contract Amount,			Clearance of Site	11.11	79
Sufficiency of the	4.11	35	Commencement of Works	8.1	59
Access after Taking Over,			Completion of Outstanding		
Right of	11.7	78	Work and Remedying		
Access Route	4.15	38	Defects	11.1	75
Access to the Site, Right of	2.1	15	Completion, Statement on	14.10	105
Adjustments for Changes			Completion, Time for	8.2	59
in Cost	13.7	94	Conditions, Unforeseeable		
Adjustments for Changes			Physical	4.12	35
in Laws	13.6	93	Confidentiality	1.12	11
Advance Payment	14.2	97	Contract Termination	1.16	13
Advance Warning	8.4	60	Contract Price, The	14.1	97
Agreement, Contract	1.6	8	Contractor to Search	11.8	78
Agreement or Determination	3.7	21	Contractor's Claims	20.1	134
Amicable Settlement	21.5	143	Contractor's Documents	4.4	29
Arbitration	21.6	143	Contractor's Documents,		
Archaeological and			Employer's Use of	1.10	11
Geological Findings	4.23	43	Contractor's Equipment	4.17	40
Assignment	1.7	9	Contractor's General		
Assistance	2.2	16	Obligations	4.1	27
Authorities, Delays			Contractor's Obligations after		
Caused by	8.6	62	Termination	16.3	117
Avoidance of Interference	4.14	38	Contractor's Obligations:		
Care of the Works,			Tests on Completion	9.1	69
Responsibility for	17.1	119	Contractor's Operations		
Care of the Works,			on Site	4.22	42
Liability for	17.2	119	Contractor's Personnel	6.9	51
Certificate, Performance	11.9	79	Contractor's Records	6.10	52
Claims	20.1	134	Contractor's Representative	4.3	29
Claims for Payment and/or			Contractor's Superintendence	6.8	51
EOT	20.2	134	Cooperation	4.6	31

Guide to the FIDIC Conditions of Contract for Construction: The Red Book 2017, First Edition.
Michael D. Robinson.
© 2023 John Wiley & Sons Ltd. Published 2023 by John Wiley & Sons Ltd.

	Sub-Clause	Page		Sub-Clause	Page
Cost, Adjustments for			Documents, Employer's		
Changes in	13.7	94	Use of Contractor's	1.10	11
Currencies of Payment	14.15	107	Documents, Priority of	1.5	8
DAAB – see Dispute			Duty to Minimise Delay	18.3	124
Avoidance/			Employer-Supplied Materials		
Adjudication Board			and Employer's		
Daywork	13.5	92	Equipment	2.6	18
Defective Work off Site,			Employer's Documents,		
Remedying of	11.5	77	Contractor's Use of	1.11	11
Defects, Failure to Remedy	11.4	77	Employer's Financial		
Defects and Rejection	7.5	55	Arrangements	2.4	17
Defects Notification Period,			Employer's Liability,		
Extension of	11.3	77	Cessation of	14.14	107
Defects, Remedying of	11.1	75	Employer's Personnel and		
Definitions	1.1	1	Other Contractors	2.3	17
Delay Damages	8.8	63	Engineer, The	3.1	19
Delay Drawings or			Engineer, Delegation by the	3.4	20
Instructions	1.9	9	Engineer, Replacement of	3.6	21
Delays Caused by Authorities	8.6	62	Engineer's Duties and		
Discharge	14.12	106	Authority	3.2	19
Disorderly Conduct	6.11	52	Engineer's Instructions	3.5	21
Dispute Avoidance/			Engineer's Representative	3.3	20
Adjudication Board,			Environment, Protection		
Constitution of the	1.1	139	of The	4.18	40
Dispute Avoidance/			Exceptional Events	18.1	123
Adjudication			Exceptional Event, Notice		
Board's Decision,			of an	18.2	124
Failure to Comply With	21.7	144	Exceptional Event,		
Dispute Avoidance/			Consequences of an	18.4	124
Adjudication			Extension of Defects		
Board' Decision,			Notification Period	11.3	77
Obtaining	21.4	140	Extension of Time		
Dispute Avoidance/			for Completion	8.5	61
Adjudication Board,			Failure to Pass Tests		
No DAAB			On Completion	9.4	70
in Place	21.8	144	FPC, Issue of	14.13	107
Dispute Avoidance/			Goods, Transport of	4.16	39
Adjudication Board			Health and Safety Obligations	4.8	32
Member(s),			Health and Safety of		
Failure to Appoint	21.2	140	Personnel	6.7	50
Disputes, Avoidance of	21.3	140	Indemnities by Contractor	17.4	121
Documents, Care and			Indemnities by Employer	17.5	122
Supply of	1.8	9	Indemnities, Shared	17.6	122
Documents, Contractor's			Inspection (by the Employer's		
Use of Employer's	1.11	11	Personnel)	7.3	54

	Sub-Clause	Page		Sub-Clause	Page
Insurance to be Provided			Payment, Delayed	14.8	103
by the Contractor	19.2	129	Payments, Schedule of	14.4	100
Insurance: General			Performance Certificate	11.9	79
Requirements	19.1	127	Performance Security	4.2	27
Instructions, Delayed			Personnel, Key	6.12	52
Drawings or	1.9	9	Personnel, Contractor's	6.9	51
Intellectual and Industrial			Personnel and Other		
Property Rights	17.3	121	Contractors, Employer's	2.3	17
Interference, Avoidance of	4.14	38	Persons, Recruitment of	6.3	49
Interference With Tests on			Plant and Materials Intended		
Completion	10.3	73	for the Works	14.5	101
Interim Payment Certificates,			Plant and Materials,		
Application For	14.3	99	Ownership of	7.7	56
Interpretation	1.2	6	Priority of Documents	1.5	8
Issue of FPC	14.13	107	Programme	8.3	60
Issue of IPC	14.6	102	Progress, Rate of	8.7	62
Joint and Several Liability	1.14	12	Progress Reports	4.20	41
Labour, Engagement of			Provisional Sums	13.4	91
Staff and	6.1	49	Quality Management and		
Labour, Facilities for			Compliance Verification	4.9	33
Staff and	6.6	50	Records, Contractor's	6.10	52
Law and Language	1.4	7	Release from Performance		
Laws, Adjustment for			under the Law	18.6	126
Changes in	13.7	94	Remedial Work	7.6	56
Laws, Compliance With	1.13	12	Remedying Defects, Failure to	11.4	77
Laws, Labour	6.4	50	Remedying Defects	11.1	75
Liability, Cessation of			Remedying Defects, Cost of	11.2	76
Employer's	14.14	107	Replacement of the Engineer	3.6	21
Liability, Joint and Several	1.14	12	Representative, Contractor's	4.3	29
Liability, Limitation of	1.15	13	Representative, Engineer's	3.3	20
Manner of Execution	7.1	53	Responsibility for Care		
Meetings	3.8	24	of Works	17.1	119
Method of Measurement	12.2	83	Resumption of Work	8.13	66
Nominated Subcontractors	5.2	46	Retention Money,		
Notice to Correct	15.1	109	Release of	14.9	104
Notices and Other			Re-testing	9.3	70
Communications	1.3	6	Right to Vary	13.1	87
Obligations, Unfulfilled	11.10	79	Rights of Way and Facilities	4.13	38
Obligations, Contractor's			Royalties	7.8	57
General	4.1	27	Samples	7.2	53
Omissions	12.4	85	Schedule of Payments	14.4	100
Payment	14.7	103	Search, Contractor to	11.8	78
Payment after Termination			Security of the Site	4.21	42
by the Contractor	16.4	117	Security, Performance	4.2	27
Payment, Currencies of	14.15	107	Setting Out	4.7	32

Sub-Clause	Page		Sub-Clause	Page
Sub-Clause	**Page**		**Sub-Clause**	**Page**
Site, Clearance of	11.11	79	Termination for Contractor's		
Site, Contractor's			Default, Payment after	15.4	113
Operations on	4.22	42	Termination for Employer's		
Site Data, Use of	4.10	34	Convenience	15.5	113
Site Data and Items of			Termination for Employer's		
Reference	2.5	17	Convenience,		
Site, Right of Access to the	2.1	15	Valuation after	15.6	114
Site, Security of the	4.21	42	Termination for Employer's		
Staff and Labour,			Convenience,		
Engagement of	6.1	49	Payment after	15.7	114
Staff and Labour, Facilities for	6.6	50	Termination, Contractor's		
Statement on Completion	14.10	105	Obligations After	16.3	117
Statement, Final	14.11	105	Testing by the Contractor	7.4	55
Subcontractors	5.1	45	Tests after Remedying		
Subcontractors, Nominated	5.2	46	Defects, Further	11.6	78
Superintendence, Contractor's	6.8	51	Tests on Completion,		
Surfaces Requiring			Contractor's		
Reinstatement	10.4	74	Obligations	9.1	69
Suspension, Consequences of			Tests on Completion, Delayed	9.2	70
Employer's	8.10	64	Tests on Completion,		
Suspension by the Contractor	16.1	115	Failure to Pass	9.4	70
Suspension, Employer's	8.9	64	Tests on Completion,		
Suspension, Payment for			Interference with	10.3	73
Plant and Materials after			Time for Completion	8.2	59
Employer's	8.11	65	Time for Completion,		
Suspension, Prolonged	8.12	66	Extension of	8.5	61
Taking Over Parts of the			Training	4.5	31
Works	10.2	72	Transport of Goods	4.16	39
Taking Over the Works and			Unforeseeable Physical		
Sections	10.1	71	Conditions	4.12	35
Temporary Utilities	4.19	40	Unfulfilled Obligations	11.10	79
Termination, Optional	18.5	125	Valuation of the Works	12.3	94
Termination by the Contractor	16.2	116	Value Engineering	13.2	89
Termination by Contractor,			Variation Procedure	13.3	90
Payment after	16.4	117	Wages and Conditions of		
Termination for Contractor's			Labour, Rates of	6.2	49
Default	15.2	110	Working Hours	6.5	50
Termination for Contractor's			Works, Valuation of the	12.3	84
Default, Valuation after	15.3	112	Works to be Measured	12.1	81